Fish Pharmacology and Toxicology

The Author

Prof. Dr. Govind Pandey is probably "only person in Madhya Pradesh and alone veterinarian in India with maximum academic qualifications" (about 20 degrees/diplomas/certificates). His 'Biography' is included in the famous book, "Who's Who in the World 2011" (28th Edn., America). With more than 35 years of experience in 'Research/Teaching/Administration/ Extension', Prof. Pandey took BVSc and AH, MVSc and AH and PhD (Hons.) in Pharmacology and Toxicology from the College of Veterinary Science and AH (JNKVV), Jabalpur in 1978, 1980 and 1990, respectively. He has also secured MBA, LLB, LLM, MA (Soc.), MA (Hin.), MA (Eng.), MA (Pol.), Acharya (Astr.), PGDPA and LSG (Pub. Ad.), PGDCA, Sahitya Ratna, Ayurved Ratna, AIT (Art. Insem.), and PGPHT (Poul. Sci.). He is working on different aspects of Pharmacology and Toxicology (including Drug Nanotechnology, Fishery Science and other areas of Life Sciences), and Hindi literature. He has also keen interest in HRM, Political Science, Sociology, Public Administration, Law and Astrology.

Prof. Pandey has unique distinction of working with 'excellence' in government institutes. During his career started from 7th August 1980, he worked in many capacities like "Veterinary Assistant Surgeon/Senior Veterinary Surgeon/Drawing Disbursing Officer (DDO)" under AH Department, Govt. of MP; "Chief Executive Officer/Block Development Officer/DDO" of some Janapad Panchayats under PRD Department, Govt. of MP; and "Assistant Professor/Professor/ Principal Scientist and Head" of Pharmacology in Pharmacy colleges. He served as "Deputy Director/Associate Professor/Senior Scientist" of Research Services, Nanaji Deshmukh Veterinary Science University (NDVSU), Jabalpur from 20th April, 2012 till he resumed the post of "Professor/ Principal Scientist and Head", Department of Pharmacology and Toxicology, College of Veterinary Science and AH, Rewa (NDVSU, Jabalpur) on 26th November, 2012. For great contribution in science/research/Hindi literature, Dr. Pandey has received '33 Awards/Honours/Recognitions'. His PhD work on 'indigenous hepatogenic drugs' was adjudged as the outstanding research work, for which he received "Sri Ram Lal Agrawal National Award" (1991; Indian Herbs, Saharanpur, UP). His other prestigious awards in include "Eminent Scientist Gold Medal" (2015), "Award for Excellence in Research" (Academic Brilliance Award-2014); "Best Teacher and Promoter Award" (2014); "Award for Excellence"; "Best Researcher of the Year" (2014) and "Senior Research Fellowship (1986)", besides many "Best Paper Awards". He has been honoured with 3 fellowship titles, "Fellow of Academy of Sciences for Animal Welfare (FASAW)", "Fellow of Society of Life Sciences (FSLSc)" and "Fellow of International Science Congress Association (FISCA)". Dr. Pandey has guided many research scholars and carried out many projects. He has investigated some "antihepatotoxic and anticancer herbal drugs, oestrogen induced cancer model, and ipomeamarone and paracetamol induced hepatotoxic models". Research on 'oestrogen' (ethinyl oestradiol, an oral contraceptive and hormonal replacement therapy used by millions of women) induced cancer and its treatment, and 'ipomeamarone' (from injured sweet potato tuber) induced hepatotoxicity were successfully done for the first time in India. These researches were widely acclaimed by the eminent personalities, scientific societies and media.

Dr. Pandey has authored "13 Books/Manuals" and edited "2 Books/Manuals" in science. He has published more than "250 Papers", including "8 Book Chapters". He has presented over "50 Scientific Papers" in more than "40 Conferences". In Hindi literature, he has authored "5 Books"; edited "1 Book"; released 2 audiocassettes ("Tarang" and "Archan") of own lyrics; and published/broadcasted several poems/lyrics/dramas/stories through different media. Dr. Pandey has been "Editor/Mentor/Editorial Board Member/Reviewer" of many scientific and literary books/journals/papers/magazines, including "Chief Editor" of 'International Journal of Animal, Veterinary, Fishery and Allied Sciences'. He has also been the "Life Member/Office-Bearer" of about 40 scientific, research, professional, literary and cultural organizations/associations/ societies/journals. Prof. Pandey has been used as the "Chairperson/Chief Guest/Judge/Expert/ Advisor/Resource Person" in many conferences/committees/programmes of different government and non-government organizations/associations/societies. He has delivered several speeches in a number of platforms, and successfully organized many academic, official, literary, cultural and social programmes in an exemplary manner. Dr. Pandey was the "Badminton Captain, NCC Sergeant, Literary Secretary and Hostel Prefect" at the College of Veterinary Science and AH, Jabalpur, and passed "NCC B and C Certificates" and "2 years' Course of National Service Scheme".

argue that gathering experimental data in fish for these compounds may be done more efficiently by considering mammalian toxicology data. Extensive mammalian pharmacological and toxicological studies are central to development of medicines and these can provide valuable information to guide ecotoxicological studies. For β-blockers, we can increasingly exploit the knowledge from molecular approaches to understand phenotypes and functions of adrenergic receptors in mammals versus fish. Some β-adrenergic receptors have been characterized in fish using both traditional molecular cloning methods, or via mining of genomic sequences from various organisms. These approaches demonstrate that fish have β-adrenergic receptors very similar to those present in mammals. Since we believe that any effects of β-blockers in fish are most likely to be mediated via β-adrenergic receptors, it is the physiological processes regulated by these receptors that are most likely to be affected. Thus, cardiovascular dysfunction is one possible consequence of exposure of fish to these compounds, leading to impaired fitness (*e.g.*, reduced growth and fecundity). More broadly, conceptual mathematical models suggest it might be possible to predict plasma concentrations of β-blockers in fish from mammals, although these models cannot be regarded as reliable until thoroughly validated. Experimental data are therefore urgently needed to define plasma levels and metabolism of β-blockers compared in fish with mammals. Finally, accurate data are essential for pharmaceuticals in order to compare nominal concentration data in terms of either the drug free base or the drug salt complex.

'Toxicology of Fish'

☆ Covers interaction of chemical contaminants with freshwater and marine fishes.

☆ Explores the interplay of feedback between controlled laboratory experiments and field investigations.

☆ Provides in-depth coverage which ranges from molecular level to community and population levels.

☆ Presents case studies which give insight into the effects of chemical classes perturbing aquatic systems.

Therefore, '*Toxicology of Fish*' covers toxic responses ranging from reduced reproduction and/or abnormal development, growth and differentiation. '*General principles*' of '*Toxicology of Fish*' describe bioavailability of chemicals present in the aquatic environment to fish, processes governing chemical distribution within them, how fish metabolize organic chemicals and fundamental mechanisms of chemical or drug toxicity. The '*key target systems and organismal effects*' of fish **toxicology** describe the key target organ systems for chemical impacts in fish, how chemicals produce cancer in these animals and how fish can develop resistance to chemical or drug toxicity. The '*methodologies and applications*' covers the methods for assessment of chemical effects on the fish like toxicity tests, biomarkers, simulated ecosystems and modeling approaches, and the use of data from such studies in ecological risk assessments. The '*case studies*' provide examples of how the principles and approaches are actually useful. Illustrated by case studies of actual, large-scale field investigations, the *'Toxicology of Fish'* assesses the untoward effects in laboratory models and wild fish. This subject represents the detailed information on the bioavailability of chemical pollutants, their distribution, metabolism and excretion in the host fish, and their mechanisms and sites of toxic responses.

Some Terms in Fish Pharmacology and Toxicology

1. *Application factor (AF):* It is empirically derived *'safe'* concentration of a chemical.

2. *Baseline toxicity:* It refers to narcosis which is a depression in biological activity due to toxicants being present in the organism.

3. *Bioconcentration factor (BCF):* It is the ratio of average chemical concentration in the tissues of organism under steady-state conditions to the average chemical concentration measured in the water to which organisms are exposed.

4. *Biomagnification:* This is the process by which the concentration of a chemical in the tissues of an organism increases as it passes through several levels in the food web.

5. *Biomonitoring:* The consistent use of living organisms to analyze environmental changes over time.

6. *Critical body residue (CBR):* This is an approach which routinely examines whole body chemical concentrations of an exposed organism that is associated with an adverse biological response.

7. *Effluent:* This is liquid, industrial discharge which contains chemical toxicants.

8. *KOW:* It is the octanol-water partition coefficient which represents the ratio of the concentration of octanol to the concentration of chemical in the water.

9. *Lowest observed effect concentration (LOEC):* The lowest test concentration which has a statistically significant effect over a specified exposure time is the LOEC.

10. *Maximum acceptable toxicant concentration (MATC):* An estimated value which represents highest *'no-effect'* concentration of a specific substance within the range including the *'no observed effect concentration'* (NOEC) and LOEC is called MATC.

11. *Mechanism of action:* This is the detailed events which take place at the molecular level during an adverse biological response.

12. *Median effective concentration (EC_{50}):* This is the chemical concentration which is expected to have one or more specified effects in 50 per cent of a group of organisms.

13. *Median lethal concentration (LC_{50}):* This is the chemical concentration which is expected to kill 50 per cent of a group of organisms.

14. *Mode of action:* This is a set of common behavioural or physiological sign which represents a type of adverse response.

15. *No observed effect concentration* (*NOEC*): This is the highest test concentration for which no effect is observed relative to a control over a specified exposure time.

16. *Quantitative structure-activity relationship* (*QSAR*): It is a method of modeling the relationship between biological activity and the structure of organic chemicals.

Environmental Toxicology

Definition, History and Scope

'*Environmental Toxicology*' (also called as '*Entox*') is a multidisciplinary field of science concerned with the study of '*harmful effects*' of different chemical, biological and physical agents on living organisms. However, '*Ecotoxicology*' is a branch of environmental toxicology that deals with the studying of harmful effects of '*toxicants*' at the population and ecosystem levels.

Rachel Carson is considered as the '*mother*' of environmental toxicology, because she made it a distinct field within toxicology in the year 1962 with the publication of her book, '*The Silent Spring*', which covered the effects of uncontrolled pesticide use.

The organisms are exposed to toxicants at different stages of their life cycle. The degree of toxicity may vary depending on where the organism is found within its food circle. The '*bioaccumulation*' occurs when the molecular compounds are stored in the fatty tissues of an organism. Over time, this leads to the establishment of a '*trophic cascade*' and the '*biomagnifications*' of specific toxicants. The '*biodegradation*' releases carbon dioxide (CO_2) and water as byproducts into the environment. This process is typically limited in areas affected by environmental toxicants. The harmful effects of chemical and biological agents can include toxicants from '*pollutants, pesticides and fertilizers*', all of which can impact an organism and its community through shifts in species diversity and abundance. Thus the changes in population dynamics impact the ecosystem by altering its productivity and stability. Legislation has been implemented since the early 1970s to ensure that harmful effects of environmental toxicants are minimized for all species. Unfortunately, we are facing the risk of entering in a '*dark age*' due to longstanding limitations in the implementation of simple conceptual modes.

Environmental Toxicants and their Toxicity

There are several sources of environmental toxicity which can lead to the presence of toxicants in our food, water and air. These sources may be organic and inorganic pollutants, pesticides and biological agents, all of which can have harmful effects on the living organisms. Some of these toxicants or pollutants are as follows:

1. Heavy Metals

They are found in food sources, *e.g.*, fish can also have harmful effects. These metals include lead (Pb), mercury (Hg), cadmium (Cd) and aluminium (Al), etc. The rainbow trout fish are exposed to higher Cd levels and grow at a slower rate than fish exposed to lower levels or none. Moreover, Cd can potentially alter the productivity and mating behaviours of these fishes. The heavy metals cannot only affect the behaviours, but also the genetic makeup in the aquatic organisms. A study examined the genetic diversity in wild yellow perch along various heavy metal concentration gradients in lakes polluted by mining operations. Along the gradient, the genetic diversity over all loci was negatively correlated with liver Cd contamination. In addition, there was a negative correlation seen between copper (Cu) contamination and genetic diversity. Some aquatic species have evolved heavy metal tolerances. In response to high heavy metal concentrations, *Chironomus riparius* (a Dipteran fish species of the midge family, *Chironomidae*) has evolved to become tolerant to Cd toxicity in aquatic environments. Altered life histories, increased Cd excretion and sustained growth under Cd exposure are the evidences which show that *C. riparius* exhibits genetically based heavy metal tolerance.

2. Polychlorinated Biphenyls (PCBs)

These organic pollutants are still present in our environment today despite being banned in many countries like USA and Canada. Due to the persistent nature of PCBs in aquatic ecosystems, many aquatic species contain high levels of this chemical; *e.g.*, fish farmed salmon have been shown to have significantly higher PCB levels.

3. Pesticides

They are a major source of environmental toxicity. These chemically synthesized agents have been known to persist in the environment long after their administration. The poor biodegradability of pesticides can result in bioaccumulation of chemicals in different organisms along with biomagnification within a food web. Pesticides can be categorized according to the pests they target. For example, the insecticides are used to eliminate the agricultural pests that attack various fruits and crops; however, the herbicides target herbal pests like weeds and other unwanted plants that reduce crop production. Dichloro-diphenyl-trichloroethane (DDT) is an organochlorine insecticide which has been banned due to its adverse effects on both humans and wildlife. The insecticidal properties of DDT were first discovered in 1939. Following this discovery, DDT was widely used by farmers in order to kill the agricultural pests, *e.g.*, potato beetle, coddling moth and corn earworm. In 1962, the harmful effects of the widespread and uncontrolled use of DDT were detailed by Rachel Carson in her book, *'The Silent Spring'*. Such large quantities of DDT and its metabolite, dichlorodiphenyl-dichloro-ethylene (DDE) released into the environment, were toxic to both animals and humans. The DDT is not easily biodegradable and, thus the chemical accumulates in soil and sediment runoff. The water systems become polluted and the marine life such as fish and shellfish accumulate DDT in their tissues. Further, this effect is amplified when animals who consume the fish also consume the chemical, demonstrating the biomagnification within the food web. The process of biomagnification has detrimental effects on different bird species because the DDT and DDE accumulate in their tissues inducing egg-shell thinning. The rapid declines in bird populations have been seen in Europe and North America as a result. The humans who consume animals or plants that are contaminated with DDT experience adverse health effects. The studies have shown that DDT has damaging effects on the liver, nervous system and reproductive system of humans. By 1972, the United States

'*Environmental Protection Agency*' (EPA) banned the use of DDT in USA. Despite the regulation of this pesticide in North America, it is still used in certain areas of the world. Traces of this chemical have been found in noticeable amounts in a tributary of the Yangtze river in China, indicating that this pesticide is still used in this area. Sulfuryl fluoride is an insecticide which is broken down into fluoride and sulphate when released into the environment. Fluoride has been known to negatively affect the aquatic wildlife. The elevated levels of fluoride have been proven to impair the feeding efficiency and growth of the common carp fish (*Cyprinus carpio*). Exposure to fluoride alters the ion balance, total protein and lipid levels within these fishes that changes their body composition and disrupts different biochemical processes.

Aquatic Toxicology

Definition, History and Importance

'*Aquatic Toxicology*' is the study of the effects of manufactured chemicals and other anthropogenic and natural materials and activities on aquatic organisms at various levels of organization, from subcellular through individual organisms to communities and ecosystems. It is a multidisciplinary field which integrates toxicology, aquatic ecology and aquatic chemistry. This field of study includes freshwater, marine water and sediment environments. The common tests include standardized acute and chronic toxicity tests lasting between 24 and 96 hr (for acute test) to 7 days or more (for chronic test). These tests measure endpoints like survival, growth, reproduction, that are measured at each concentration in a gradient, along with a control test.

While basic research in toxicology began in multiple countries in the 1800s, it was not until around the 1930s that the use of '*acute toxicity*' testing, especially on fish, was established. Over the next two decades, the effects of chemicals and wastes on non-human species became more of a public issue and the era of the '*pickle-far bioassays*' began as efforts increased to standardize toxicity testing techniques. In the United States of America (USA), the passage of the '*Federal Water Pollution Control Act*' of 1947 marked the first comprehensive legislation for the control of water pollution, and was followed by the '*Federal Water Pollution Control Act*' in 1956. In 1962, the public and governmental interests were renewed, in large part due to the publication of Rachel Carson's book, '*The Silent Spring*', and thee years later the '*Water Quality Act*' was passed which directed States to develop water quality standards. Public awareness, as well as scientific and governmental concern, continued to grow throughout the 1970s and by the end of the decade research had expanded to include hazard evaluation and risk analysis. In the subsequent decades, aquatic toxicology has continued to expand and internationalize so that there is now a strong application of toxicity testing for environmental protection.

In the USA, the aquatic toxicology plays an important role in the wastewater permit programmes. In addition to analytical testing for known pollutants, the aquatic, whole effluent toxicity tests have been standardized and are performed routinely as a tool for evaluating the potential harmful effects of effluents discharged into the surface waters. For '*Clean Water Act*' under the US EPA, there are water quality criteria and water quality standards derived from the aquatic toxicity tests.

Aquatic Toxicology Tests

The aquatic toxicology or toxicity tests (assays) are used to provide qualitative and quantitative data on adverse (deleterious) effects on organisms from a toxicant. Toxicity tests can be used to assess the potential for damage to an aquatic environment and provide a database that can be used to assess the risk associated within a situation for a specific toxicant.

The aquatic toxicology tests can be performed in the field or laboratory. The field experiments generally refer to multiple species exposure and laboratory experiments generally refer to single species exposure. A dose response relationship is most commonly used with a sigmoidal curve to quantify the toxic effects at a selected endpoint or criteria for effect (*i.e.*, death or other adverse effect to the organism). Concentration is on the *'x-axis'* and percent inhibition or response is on the *'y-axis'*. The type of test used depends on many factors, such as specific regulatory agency conducting the test, resources available, physical and chemical characteristics of the environment, type of toxicant, test species available, laboratory versus field testing, endpoint selection, and the time and resources available to conduct the assays.

Criteria for effects, or endpoints tested for, include lethal and sublethal effects. Various types of toxicity tests can be done on different test species. Various species differ in their susceptibility to chemicals, most likely due to differences in accessibility, metabolic rate, excretion rate, genetic factors, dietary factors, age, sex, health and stress level of the organism. Common standard test species are daphnids (*Daphnia magna, D. pulex, D. pulicata, Ceriodaphnia dubia*), fathered minnow (*Pimephales promelas*), sheepshead minnow (*Cyprinodon variegatus*), midge (*Chironomus tentans, C. ruparius*), rainbow trout (*Oncorhynchus mykiss*), mysids (*Mysidopsis*), mussels (*Mytilus*), oyster (*Crassotreas*), scud (*Hyalalla azteca*) and grass shrimp (*Palaemonetes pugio*). These species are routinely selected on the basis of availability, commercial, recreational, ecological importance, past successful use and regulatory use.

A variety of acceptable standardized test methods have been published. Some of the more widely accepted agencies to publish methods are: American Public Health Association; US EPA; American Society for Testing and Materials; International Organization for Standardization; Environment Canada; and Organization for Economic Cooperation and Development. Standardized tests offer the ability to compare results between laboratories.

Exposure Systems

There are four general techniques of exposure systems through which the controls and test organisms are exposed to deal with treated and diluted water or the test solutions.

(a) *Static test:* It exposes the organism in still water. The toxicant is added to water in order to obtain the correct concentrations to be tested. The control and test organisms are placed in the test solutions and the water is not changed for entirety of the test.

(b) *Recirculation test:* This test exposes the organism to the toxicant in a similar manner as the static test, except that the test solutions are pumped through an apparatus (*i.e.,* filter) to maintain water quality, but not reduce the concentration of the toxicant in the water. The water is circulated through the test chamber continuously, similar to an aerated fish tank. This type of test is expensive and it is unclear whether or not the filter or aerator has an effect on the toxicant.

(c) *Renewal test:* This test also exposes the organism to the toxicant in a similar manner as the static test, because it is in still water. However, in a renewal test, the test solution is renewed periodically (constant intervals) by transferring the organism to a fresh test chamber with the same concentration of the toxicant.

(d) *Flow-through test:* This test exposes the organism to the toxicant with a flow into the test chambers and then out of the test chambers. The once-through flow can either be intermittent or continuous. A stock solution of the correct concentrations of contaminant must be prepared earlier. Metering pumps or diluters will control the flow and the volume of test solution, and the proper proportions of water and contaminant will be mixed.

Different Aquatic Toxicity Tests

1. Acute Tests

They are short-term exposure tests (hours or days), and normally use lethality as an endpoint. In acute exposures, organisms come into contact with higher doses of the toxicant in a single event or in multiple events over a short period of time and usually produce immediate effects, depending on absorption time of the toxicant. These tests are normally conducted on organisms during a specific time period of the organism's life cycle, and are considered partial life cycle tests. The acute tests are not valid if mortality in the control sample is greater than 10 per cent. The results are reported in EC_{50} or concentration which will affect 50 per cent of the sample size.

2. Chronic Tests

They are long-term exposure tests (weeks, months or years), relative to the test organism's life span (>10 per cent of life span), and normally use sublethal endpoints. In chronic exposures, the organisms come into contact with low, continuous doses of a toxicant. Chronic exposures may induce effects to acute exposure, but can also result in effects which develop slowly. The chronic tests are normally considered full life cycle tests and cover an entire generation time or reproductive life cycle ('*egg to egg*'). These tests are not considered valid if mortality in the control sample is greater than 20 per cent. These results are normally reported in NOECs and LOECs.

3. Early Life Stage Tests

These are considered as subchronic exposures that are less than a complete reproductive life cycle and include exposure during early, sensitive life stages of an organism. These exposures are also called '*critical life stage, embryo-larval, or egg-fry tests*'. Early life stage tests are not considered valid if mortality in the control sample is greater than 30 per cent.

4. Short-term Sublethal Tests

They are used to evaluate the toxicity of effluents to aquatic organisms. These methods are developed by the EPA, and only focus on the most sensitive life stages. Endpoints for these tests include changes in growth, reproduction and survival. NOEC, LOEC and EC_{50} are reported in these tests.

5. Bioaccumulation Tests

These are toxicity tests which can be used for hydrophobic chemicals that may be accumulated in the fatty tissue of aquatic organisms. Toxicants with low solubilities in water usually can be stored in the fatty tissue due to high lipid content in this tissue. The storage of these toxicants within the organism may lead to cumulative toxicity. The bioaccumulation tests use BCF to predict the concentrations of hydrophobic contaminants in organisms. The BCF is the ratio of the average concentration of test chemical accumulated in the tissue of the test organism (under steady state conditions) to the average measured concentration in the water.

6. Freshwater Tests and Saltwater Tests

These tests have different standard methods, especially as set by the regulatory agencies. However, these tests normally include a control (negative and/or positive), a geometric dilution series or other appropriate logarithmic dilution series, test chambers and equal numbers of replicates and a test organism. Exact exposure time and test duration will depend on type of test (acute vs. chronic) and organism type. Temperature, water quality parameters and light will depend on regulator requirements and organism type.

7. Effluent Toxicity Tests

These are the tests done under the *'Clean Water Act, National Pollutant Discharge Elimination System (NPDES) permit programme'*, and are used by dischargers of contaminated effluent to monitor the quality of effluent into receiving waters. *'Acute effluent toxicity tests'* are used to monitor the quality of industrial effluent monthly using acute toxicity tests. The effluent is used to do the static-acute multiconcentration toxicity tests with *C. dubia* and *Pimephales promelas*. The test organisms are exposed for 48 hr under static conditions with five concentrations of the effluent. *'Short-term chronic effluent toxicity tests'* are used to monitor the quality of municipal wastewater treatment plants effluent quarterly using short-term chronic toxicity tests. The aim of this test is to ensure that the wastewater is not chronically toxic. The major deviation in the short-term chronic effluent toxicity tests and the acute effluent toxicity tests is that the short-term chronic test lasts for 7 days and the acute test lasts for 48 hr.

8. Sediment Tests

At some points, most chemicals originating from both anthropogenic and natural sources accumulate in sediment. For this reason, sediment toxicity can play a major role in the adverse biological effects seen in the aquatic organisms, especially those inhabiting benthic habitats. A recommended approach for sediment testing is to apply the *'sediment quality triad'* (SQT) which involves simultaneously examining sediment chemistry, toxicity and field alterations, so that more complete information can be gathered. Collection, handling and storage of sediment can have an effect on bioavailability, and for this reason the standard methods have been developed to suit this purpose.

Effects of Toxicants

'Toxicity' can be divided into two broad categories: direct and indirect toxicity. *'Direct toxicity'* results from a *'toxicant'* acting at the site of action in or on the organism. *'Indirect toxicity'* occurs with a change in physical, chemical or biological environment.

'Lethality' is most common effect of a toxicant, and used as an endpoint for acute toxicity tests. While conducting chronic toxicity tests, the sublethal effects are endpoints that are looked at. These endpoints include behavioural, physiological, biochemical, and histological changes.

There are a number of effects which occur when an organism is simultaneously exposed to two or more toxicants. These effects are additive, synergistic, potentiating and antagonistic effects. An *'additive effect'* occurs when combined effect is equal to a combination or sum of the individual effects. A *'synergistic effect'* occurs when the combination of effects is much greater than the two individual effects added together. *'Potentiating effect'* is an effect which occurs when an individual chemical of no effect is added to a toxicant and the combination has a greater effect than just the toxicant alone; whereas, an *'antagonistic effect'* occurs when a combination of chemicals has less of an effect than the sum of their individual effects.

Gross Appearance of Poisoned or Polluted Fishes

A variety of chemicals, toxicants, environmental pollutants or poisons can produce many signs and symptoms in fish which can be observed macroscopic (grossly by naked eyes) and microscopic examinations. To understand the location of damaged/diseased part of an affected fish, we must know the external anatomy (Figure 1) and internal anatomy (Figure 2) of a fish. Simultaneously, to know the conditions or changes in the body of poisoned or affected fish, here are some images (Figures 3–18). All these images (Figures 1–18) have been extracted from different internet websites (particularly *'google.com'*) that are duly acknowledged.

Overviews on the Pharmacology and Toxicology Research in Fish

To recognize the differences in the physiology of fish, it is necessary to understand the problems associated with drug approval for the aquatic environment. Various experiments and drugs used in fish have been elucidated in the literatures. Antibacterial, antiparasitic and anaesthetic drugs, besides pharmacokinetic and pharmacodynamic parameters have been experimented on the fish. Many groups of drugs, *e.g.*, penicillins, tetracyclines, macrolides, quinolones, sulphonamides, immunostimulants, anticancer agents, herbal drugs and vaccines, have been successfully experimented or used in fish.

Hence, fish is also used as a new model organism for various experimental studies of pharmacology and toxicology (Pandey, 2011a; Pandey *et al.*, 2013a). The potential for the application of research findings to both human and environmental

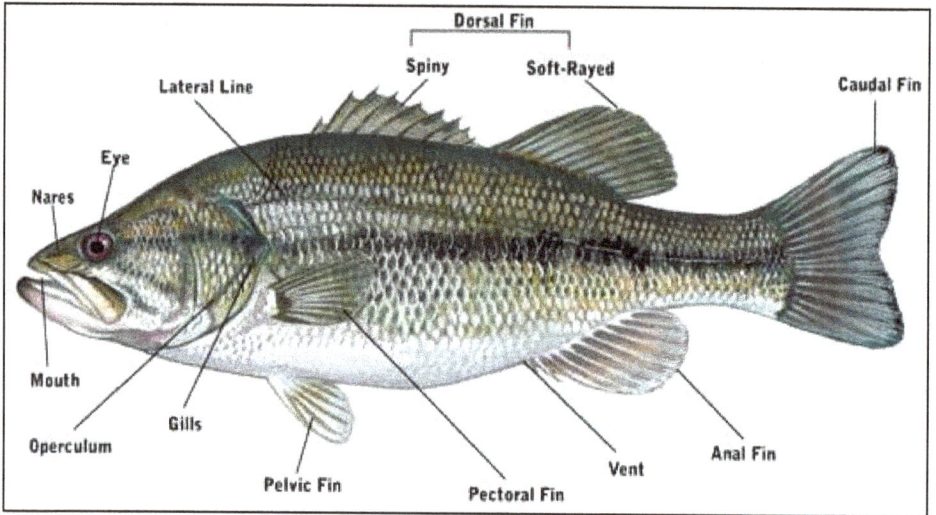

Figure 1: External Anatomy of Fish.

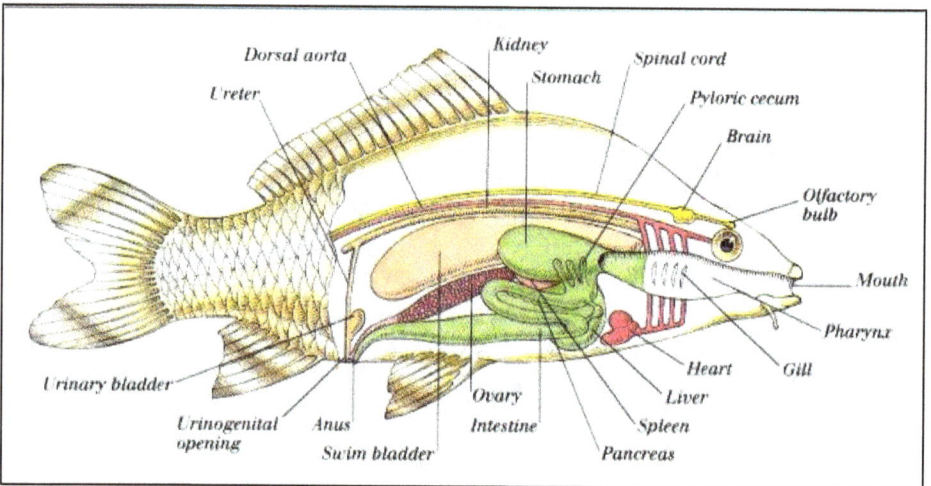

Figure 2: Internal Anatomy of Fish.

Figure 3: Poisoned Fish.

Figure 4: Poisoned Fish.

Figure 5: Poisoned Fish.

Figure 6: Poisoned Fish.

Figure 7: Poisoned Fish.

Figure 8: Poisoned Fish.

Figure 9: Poisoned Fish.

Figure 10: Poisoned Fish.

Figure 11: Poisoned Fish.

Figure 12: Poisoned Fish.

Figure 13: Poisoned Fish.

Figure 14: Poisoned Fish.

Figure 15: Poisoned Fish.

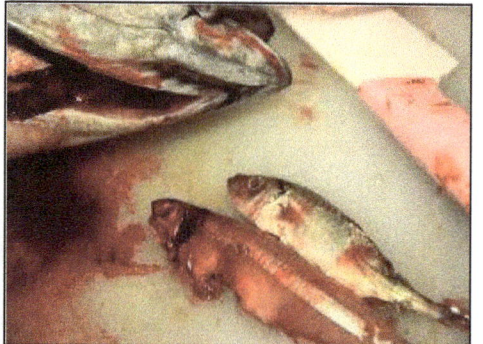

Figure 16: Poisoned Fish.

Figure 17: Poisoned Fish.

Figure 18: Poisoned Fish.

health issues makes fish species attractive and valuable alternative models in different diseases, including cancer, and pharmacological and toxicological research (Pandey, 2011a).

Fish not only play an important role for food demand of humans but they are also widely used in various biological experiments (Pandey *et al.*, 2012b and 2012f). Many indigenous and exotic fish species are found either naturally or cultured in the reservoirs (Pandey *et al.*, 2012f). One of the very popular freshwater tropical fish, *Danio rerio* (zebrafish or zebra fish) has emerged as a major model organism for biomedical research, especially in developmental genetics, neurophysiology, oncology and biomedicine (Pandey, 2011a and 2011b; Pandey *et al.*, 2012f). Like most other biomedical model organisms, the zebrafish is chosen for particular traits that make it convenient for laboratory study. The greatest advantage of zebrafish as a model system comes from its well-characterized genetics, genetic and developmental techniques and tools, and the availability of well-characterized mutants. Induction of experimental diseases, including cancer using zebrafish as a model organism may be a new and an important field of research (Madhuri, 2012; Madhuri *et al.*, 2012h; Pandey *et al.*, 2012f and 2013a).

Zebrafish is a small, robust fish, and so large numbers can be kept easily and cheaply in laboratory. It is a popular aquarium fish, frequently sold under the trade name *'zebra danio'*. They have luminous bluish-black and silvery-gold horizontal stripes or dark-blue and silvery longitudinal stripes. The transgenic zebrafish is commercially available under the trade name *'glowfish'*. Zebrafish originate from eastern Asia, and distributed throughout the south and south-east Asia (Madhuri, 2012). In fact, zebrafish has been recognized as a promising animal model for cancer research. Compared to mammals, the induction of cancer in fish appears to be easy. At the same time, zebrafish has been found to be a responsive, cost-effective lower vertebrate model system to study the mechanism of carcinogenesis. Liver has been found to be the most commonly targeted organ in the cancer studies with zebrafish. The cancers (neoplasms) which can be seen in zebrafish include seminoma, hepatocellular carcinoma (HCC), hepatocellular adenoma (HCA), adenoma of exocrine pancreas, intestinal adenocarcinoma, ultimobranchial neoplasm, thyroid neoplasm, spindle cell sarcoma and hemangioma, etc. (Madhuri *et al.*, 2012h).

Several experiments and the use of drugs have been performed in the fish (Madhuri *et al.*, 2012b; Pandey *et al.*, 2012b). Besides being used as model organism in biological experiments, fish has also been used in environmental toxicology. The aquatic animals are being engineered to increase aquaculture production, for medical and industrial research, and for ornamental reasons. Fish have been found to play an important role in assessing potential risks associated with exposure to toxic substances in aquatic environment. Hence, it has been thought that the development of transgenic fish can enhance the use of fish in environmental toxicology. India has developed experimental transgenics of rohu fish, zebra fish, cat fish and singhi fish. Genes, promoters and vectors of indigenous origin are now available for only two species namely rohu and singhi for engineering growth. Development of fish model carrying identical transgenes to those found in rodents is beneficial, and has shown that several aspects of *in vivo* mutagenesis are similar between the two classes of vertebrates. Fish shows the frequencies of spontaneous mutations similar to rodents and respond to mutagen exposure consistent with known mutagenic mechanisms. The feasibility of *in vivo* mutation analysis using transgenic fish has been shown and the potential value of transgenic fish as a comparative animal model has been illustrated. Thus, the transgenic fish can give significant contribution to study the environmental toxicity in animals (Madhuri *et al.*, 2012b).

The fish can carry pathogens and parasites. Disease in fish can lead to fish mortality, especially when the fish are young. Pathogens which can cause fish diseases include viral, bacterial, fungal, protozoan, water mould infections, etc. Fish are also exposed from different environmental pollutants, including drugs and chemicals. Most common fish diseases, particularly in freshwater aquaria, are columnaris, gill disease, ick (ich), dropsy, tail and fin-rot, fungal infections, white spot disease, pop-eye, cloudy eye, swim bladder disease, lice and nematode worms infestation, water quality induced diseases, constipation, anorexia, chilodonella, ergasilus, tuberculosis, glugea, henneguya, hexamita, hole-in-the-head disease, injuries, leeches in aquaria, lymphocystis, marine velvet, and neon-tetra disease, etc. Plants are used as traditional medicine since time immemorial to control bacterial, viral, fungal and other diseases (Madhuri *et al.*, 2012i). However, the antibiotics are frequently used to control bacterial fish diseases, but there is risk of developing antibiotic resistant strains of bacteria (Madhuri *et al.*, 2012i; Pandey *et al.*, 2012b). Diseases in fish caused by bacteria are most widespread. The herbs contain many immunologically active components such as polysaccharides, organic acids, alkaloids, glycosides and volatile oils, which can enhance immune functions. Recently, there has been increased interest in the immune stimulating function of some herbs in aquaculture (Pandey *et al.*, 2012b).

Parasitic infection may cause severe morbidity and mortality in fish. For example, flagellates of *Hexamita* parasite are often associated with high mortality in fish. The flagellate parasites interfere with nutrition by competing for essential nutrients and/ or by damaging the intestinal epithelium. 'Hexamitosis' is probably the most frequent internal flagellate parasitosis of fish, notably in young salmonids. *H. salmonis* infected fish pass both trophozoites and cysts in faeces. Many drugs have been investigated for treatment of parasite infection in different species of fish. However, the only oral pharmacological treatment of *H. salmonis* determined to date is the drugs of nitroimidazole group showing the activity against different protozoan groups, including flagellates and ciliates. Current treatment of choice is dimetridazole or metronidazole mixed in the feed. *H. salmonis* infection in the rainbow trout (*O. mykiss*) fish has been completely eradicated not only by metronidazole, but also by benznidazole, ronidazole and secnidazole. Four nitroimidazoles, *e.g.*, albendazole, aminosidine, diethylcarbamazine and nitroscanate have completely eliminated the parasite infection in fish. Three drugs have also been recommended for the treatment of protozoan parasitoses of fish, *viz.*, amprolium, bithionol and toltrazuril. Among these three (amprolium, bithionol and toltrazuril), amprolium (by oral administration) has been found effective; indeed, its oral administration is recommended for myxosporidiosis. Bithionol and toltrazuril are effective only with bath treatment (Madhuri and Pandey, 2014; Pandey, 2013a).

Ornamental fish are important commercial components of fisheries. The ornamental fish are grouped mainly into two categories, *i.e.*, egg-layers (oviparous) and live-bearers (viviparous; *e.g.*, guppies, platys, mollies and sword-tail). Zebra fish (egg-layers group) is the ornamental fish of Indian origin which is bred and reared very easily. The diseases in ornamental fish may occur due to poor water quality, temperature fluctuation, poor quality feed, etc. Common diseases like columnaris disease, fin rot and tail rot disease, and white spot disease are seen. The pop-eye disease, velvet disease, slimy skin disease, dropsy, etc. are also noticed in various aquarium fishes. The ectoparasite infection due to fluke attack, fish lice and anchor worm are seen in many cases. Potassium permanganate ($KMnO_4$) is an oxidizer which can be used to *'disinfect'* the external surfaces of fish. Correct use of $KMnO_4$ can effectively control many bacterial, parasitic and fungal agents before systemic infections occur. In the ornamental trade, correct use of $KMnO_4$ at the onset of an infection can also speed the movement of fish as they do not need to be held for lengthy (often 10-14 days) antibiotic treatments. The diseased ornamental fishes must be separated and given a mild *'dip treatment'* with $KMnO_4$ (0.002 per cent), common salt (3 per cent) or copper sulphate (0.05 per cent). Chloromycetin 40 mg in 5 L of water may be added in aquarium tank to prevent some bacterial diseases. The aquarium has to be thoroughly cleaned with disinfectants like $KMnO_4$, malachite green or formalin in case of high disease intensity. Most fish can be treated by prolonged immersion in a 2 mg/L $KMnO_4$ solution (water must retain a purple colour for at least 4 hr), although some species may be sensitive to it and may not tolerate a full strength (2 mg/L) bath. Because of its harsh oxidizing properties, $KMnO_4$ should not be applied to fish more frequently than once per week, otherwise the mortality may result. The $KMnO_4$ can stain skin and clothing, so care is suggested when handling it. The badly infected fishes should be taken out to prevent further spreading of diseases (Madhuri and Pandey, 2014; Pandey, 2013b).

The parasitic outbreak in fish acts as an important limiting factor for aquaculture. Heavy infection of *Trichodina* parasites has caused gigantic financial losses (Pandey, 2013c; Pandey *et al.*, 2012a). Diseases caused by *Aeromonas hydrophila* bacterium are some of the most widespread in freshwater fish culture. Nowadays, a large portion of the world population, especially in developing countries depends on the traditional system of medicine for a variety of diseases. The medicinal plants (herbal drugs) can be used not only against diseases but even more so, as growth promoters, stress resistance boosters and preventatives of infections. Phytomedicines also provide a cheaper source for treatment and greater accuracy than chemotherapeutic agents without causing toxicity. The phytochemicals, *e.g.*, tannins, alkaloids and flavonoids present in medicinal plants may have antimicrobial activity (Pandey *et al.,* 2012a). Treatment of ectoparasitic diseases in freshwater fish with formalin seems to be ineffective; therefore, medicinal plants can solve this problem. Treatment with medicinal plants having antibacterial activity is a potentially beneficial alternative in the aquaculture. Recently, research has been initiated to evaluate the feasibility of herbal drugs in fish diseases. The medicinal plants (herbs) can act as immunostimulants, conferring early activation to the nonspecific defense mechanisms of fish and elevating the specific immune response (Pandey, 2013c; Pandey *et al.,* 2012a, 2012b and 2012c). The nonspecific immune functions like bacteriolytic activity and leukocyte function of fish have been improved by some herbs (Madhuri *et al.,* 2012i; Pandey, 2013c; Pandey *et al.,* 2012a and 2012b). Garlic (*Allium sativum*) and Indian almond (*Terminalia catappa*) have been used to treat tilapia fish ectoparasites, *viz., Trichodina* species. Since medicinal plants stimulate only nonspecific immune responses, vaccines may be a better way to prevent the fish diseases. These plants may be used as vaccine adjuvants (Madhuri and Pandey, 2014; Pandey, 2013c).

Aquaculture has been a growing activity for more than 20 years worldwide. The bacterial infections are the major cause of mortality in aquaculture. Common fish pathogenic bacteria, *e.g.*, *Streptococcus agalactiae*, *Lactococcus garvieae*, *Enterococcus faecalis* (all Gram-positive), *A. hydrophila* and *Yersinia ruckeri* (both Gram-negative) can cause infectious diseases. *A. hydrophila*, the most common bacterial pathogen in freshwater fish, is the aetiological agent of many pathological conditions, including tail rot, motile Aeromonas septicemia and epizootic ulcerative syndrome as a primary pathogen. Continuous use of antibacterial agents in aquaculture has resulted into resistant bacterial strains in the aquatic environment. Treatment of bacterial diseases with different herbs has been safely used in organic agriculture, veterinary and human medicine. Treatment with medicinal plants (herbs) having antimicrobial (antibacterial and antifungal) activity is a potentially beneficial alternative in the aquaculture. These herbs mitigate many of the side effects which are associated with synthetic antimicrobials. Additionally, the plant-derived phytomedicines provide a cheaper source for treatment and greater accuracy than chemotherapeutic agents. Plants have been used as traditional medicine since time immemorial to control bacterial, viral and fungal diseases. In India, 500 medicinal plant species are used against pathogenic bacteria. Recently, research has been initiated to evaluate the feasibility of herbal drugs in fish diseases. Because of the growing bacterial resistance against commercial standard and reserve antibiotics, the search for new active substances with antibacterial activity against pathogenic bacteria is of increasing importance (Madhuri *et al.,* 2012a). Fish are exposed from different chemicals, including drugs and pollutants. The most important chemicals or pollutants causing severe toxicity to fish are *'heavy metals'*, *e.g.*, arsenic (As), cadmium (Cd), copper (Cu), chromium (Cr), iron (Fe), lead (Pb), manganese (Mn), mercury (Hg), nickel (Ni), zinc (Zn), tin (Sn), etc. These chemicals have also caused cancer in fish. Besides, diethylnitrosamine (DEN), high concentrations of polycyclic aromatic hydrocarbons (PAHs) and some other metals cause the cancer of liver, skin and other organs of fish (Madhuri *et al.*, 2012e and 2012f; Pandey *et al.*, 2013a). To overcome with such situations, some medicinal plants have been found effective. They are cheaper with greater accuracy than the chemotherapeutic agents without causing toxicity (Madhuri *et al.*, 2012e).

Herbal drugs are used not only against diseases but also as growth promoters, stress resistance boosters and preventatives of infections in fishes (Madhuri *et al.*, 2012e; Pandey *et al.*, 2012c). Studies have proved that herbal additives enhance the growth of fishes and protect them from diseases. Inclusion of herbal additives in diets often provides cooperative action to various physiological functions. Beneficial role of vitamins C and E have been reported in fish nutrition, reproduction, growth and related indices. In addition, vitamins C and E are credited with modulating the stress response in fish. The biological role played by vitamins C and E is very vital for the sustained growth and health of many living organisms as well as fish. Dietary vitamins have antibody enhancement effects in fish. The herbal drugs, *viz.*, ginger, nettle and mistletoe have been used as an adjuvant therapy in rainbow trout fish through feed. The disease resistant of catla fish has been produced through the immersion herbal treatment (neem, garlic and turmeric) of spawn (Pandey *et al.*, 2012c). Small ornamental fish farms with an assortment of fish require small amounts of various diets with particular ingredients. It is not cost effective for commercial manufacturers to produce very small quantities of specialized feeds. Most feed mills will only produce custom formulations in quantities of more than one ton, and medicated feeds are usually sold in 50-pound bags. Small fish farmers, hobbyists and laboratory technicians are, therefore, left with the option of buying large quantities of expensive feed, which often goes to waste. Small quantities of fish feeds can be made quite easily in the laboratory, classroom, or at home, with common ingredients and simple kitchen or laboratory equipment. Hence, the knowledge of fish feed formulation and feeding technology related to live feed for fish larvae, fish feeds, fish feed ingredients, common fish feed stuffs, animal and plant sources of feeds for culture fish, and fish feeding methods is necessary (Madhuri *et al.*, 2012g).

Heavy metals mainly include Pb, Hg, Cd, Cr, Cu, Zn, Mn, Ni, Ag, etc. Heavy metals, *viz.*, As, Cd, Pb and Hg are most toxic to humans, animals, fishes and environment. Excessive levels of heavy metals are detrimental. They destabilize ecosystems because of their bioaccumulation in organisms, and toxic effects on biota and even death in most living beings. All heavy metals, in spite some of them are essential micronutrients, have their toxic effects on living organisms via metabolic interference and mutagenesis. The bioaccumulation of toxic metals can occur in the body and food chain. Thus, the toxic metals generally exhibit chronic toxicity. The heavy metals like Pb and Hg have significant toxic effects. Of course, the heavy metals are important pollutants for fish because they are not eliminated from aquatic systems by natural methods like organic pollutants, and are enriched in mineral organic substances. Occurrence of heavy metals differs in fishes, depending on their age, development and other physiological factors. Among animal species, fish are inhabitants that can be highly affected by these toxic pollutants. Heavy metals can damage different organs. They can enter into water via drainage, atmosphere, soil erosion and all human activities by different ways. As the heavy metals are concentrated more in the environment, they enter biogeochemical cycle, causing toxicity (Madhuri *et al.*, 2012f; Pandey, 2013e; Pandey and Madhuri, 2014). Hence, the heavy metals are the most important pollutant for aquatic animals because of their toxicity and accumulation by marine organisms. Fish may be more greatly affected by anthropogenic pollution sources. Fish are highly exposed from the heavy metals, like Hg, leading to severe toxicity, both in fish and human beings. Fish accumulate substantial concentrations of Hg in their tissues, and thus the fish are single largest source of Hg for humans through fish eating. The organic forms of Hg (*e.g.*, methyl Hg) are more toxic than the inorganic forms due to ease of absorption into the human system. Communities that relied on fish intake for daily nutrient sustenance may be at risk from chronic, high exposure to methyl Hg, as well as other persistent organic environmental pollutants. The organic Hg compounds are most toxic to central nervous system (CNS), and may also affect the kidneys and immune system. Main symptoms of Hg poisoning in humans include kidney damage, disruption of nervous system, damage to brain functions, DNA and chromosomal damage, allergic reactions, sperm damage, birth defects, and miscarriages. The greater concern for Hg exposure is not to adult human, but to developing foetus. Methyl Hg content of fish varies by the species and size of fish, as well as harvest location. The *'Food and Drug Administration'* (*FDA*) level for Hg in fish is 1 ppm. The fish with levels higher than this should probably be avoided (Madhuri and Pandey, 2014; Pandey, 2013e; Pandey *et al.*, 2012d).

Pollution of rivers and streams with chemical contaminants has become one of the most critical environmental problems. Similar to humans and other species of animals, fish have also been polluted with a variety of pollutants. Unintentionally added chemicals causing fish toxicity include organochlorine pesticides, PCBs and other persistent chemicals in feed, chemicals in construction materials, and metabolites and degradation products of intentionally added chemicals. High levels of PCBs have been reported in farmed fishes. The fishes caught in highly polluted water could cause cells of some kinds of cancers to multiply rapidly. The PCBs are persistent environmental contaminants that are ubiquitous in the environment due to intensive industrial use. Many PCB congeners persist in ambient air, water, marine sediments and soil at low levels throughout the world. Humans may be exposed to detectable quantities of PCBs when they eat fish, use fish oils in cooking, or consume meat, milk or cheese. General population is exposed to PCBs by inhaling contaminated air and ingesting contaminated water and food. The toxic responses to PCBs are dermal toxicity, immunotoxicity, carcinogenicity, and adverse effects on reproduction, development and endocrine functions. Epidemiological studies indicate that consumption of background levels of PCBs may cause slight but measurable impairments in physical growth and learning behaviour in children (Pandey *et al.*, 2012e).

Contamination of inland and surface waters and land/soil due to the release of variety of chemicals may prove toxic to all classes of living organisms. One such dangerous and toxic chemical is *'cyanide'*. Fish is very sensitive species to cyanide toxicity. Cyanide is innately toxic to fish, but many factors influence the degree of toxicity. Many industries emanate huge quantity of cyanide containing effluents. Wastewaters, especially generated from industries, *e.g.*, metal mining, electroplating, printed circuit board manufacturing, etc. are often toxic. There are reports that even cyanide concentrations as low as 0.01 to 0.1 mg/L are able to kill some sensitive animals species present in waters. The assessment of toxicity of sodium cyanide (NaCN) with reference to aquatic biota, especially fish is crucial in establishing the toxicity evaluation. The NaCN is known to be readily taken up by aquatic organisms and bioconcentration factors ranged from 130 to 4,900 for various organisms (snail, daphnia and fish). Data indicates that NaCN is toxic to *Labeo rohita* fish. Similarly, the toxic stress of pesticides has direct bearing on tissue chemical compounds. Thus, in order to protect the whole aquatic ecosystem, awareness must be given to farmers to control the agricultural pests by biological methods other than the use of cyanides/pesticides (Madhuri and Pandey, 2014; Pandey, 2013d).

Fish play an important role in assessing potential risks associated with exposure to toxic substances, particularly pesticides in aquatic environment. Although the pesticides have brought tremendous benefits to mankind by increasing food production and controlling the vectors of man and animal diseases; but they have also posed potential health hazards to fish. Many pesticides become pollutant and act as immunotoxic, leading to suppress the cell-mediated immunity. Fish kills occur when pesticides are improperly applied to or otherwise end up in bodies of water through either misapplication or drift. Hence, new approaches are needed to improve the assessment of health risks associated with exposure to pesticides or such chemical contaminants in the aquatic environments (Madhuri *et al.*, 2012d). Although the pesticides have brought tremendous benefits to mankind by increasing food production and controlling the vectors of man and animal diseases but they have also posed potential health hazards to fish. The fish blood is highly susceptible to internal and external environment fluctuations. The release of pesticide chemicals into the aquatic environment creates a potential for unintended adverse health impacts to both humans and non-target wildlife. Assessing the cumulative toxicity of pesticides has, therefore, been an enduring challenge for environmental health research as well as ecotoxicology. Cumulative toxicologic impact of pesticide contamination is of particular concern for salmon and steelhead fish population. The pesticides almost always occur in mixtures with other pesticides. Among the pesticides, organophosphate (OP) and carbamate pesticides form an important class of toxic compounds. Some OP pesticides include diazinon, malathion and chlorpyrifos; and carbamate pesticides include carbaryl and carbofuran, etc. Fish kills involving these insecticides have been documented. The organophosphate and carbamate pesticides are commonly detected in freshwater habitats may cause toxicity in different fish species. These pesticides inhibit the activity of acetylcholinesterase (AChE) and thus have potential to interfere with behaviours which can be essential for the survival of fish. The effects of individual anticholinesterase insecticides on aquatic species have been studied for decades (Madhuri and Pandey, 2014; Pandey and Pandey, 2013). Hence, pesticides are among the most important environmental pollutants because of their increasing use in agriculture. *'Nanotechnology'* has issues regarding the health which in turn pose bad reputation of nanotechnology in the society. To reduce the toxicity and side effects of drugs, *'nanoparticles'* (NPs) are now used, but they have been proved to impose risks to the health. There are many concerns with the use of NPs and even they can also enter through breathing and can damage the vital organs. The *'combustion derived nanoparticles'* (CDNPs) are important components that drive the adverse effects of environmental particulate air pollution. Toxicity caused by the newly engineered NPs such as *'fullerenes'*, *'carbon nanotubes'* (CNTs), *'quantum dots'*, etc. is a challenge of nanotechnology. In this way, the OP and carbamate NPs can cause high toxicity to fish, as well. Many methods are available for detection of OP and carbamate NPs (Pandey and Pandey, 2013).

Nowadays, cancer in fish has become a great problem. Several chemical carcinogens, environmental pollutants, radiant energy, and oncogenic viruses and some other microbes can cause genetic damage and neoplastic transformation of cells, leading to development of various cancers. Fish have gained increasing attention over the past three decades as valuable models for carcinogenesis research, wherein they have been found to suffer from various types of cancer (Madhuri *et al.*, 2012c; Pandey *et al.*, 2014). Many cancers in fish appear to be the result of exposure to different environmental pollutants/chemicals. Compared to mammals, cancer induction in fish appears to be easy. Liver has been found to be the most targeted organ for fish cancer. High frequencies of liver and skin cancers in brown bullheads are associated with high concentrations of PAHs and some metals in the sediment. The HCC in sauger and walleye are associated with heavy loadings of extremely fine particulates, which were produced when copper *'stamp sands'* were reprocessed. Tumours/cancers in zebrafish are more common. The beluga whales are getting cancer, while those in the less polluted water do not suffer from cancer (Madhuri *et al.*, 2012c; Pandey *et al.*, 2013a, 2013b and 2014). Fish living in a polluted water reservoir use the contaminated water to rinse their gills; this results in the deposition of PAHs in the fish body. Contamination of foodstuffs by heavy metals such as As, Cd, Cr, Ni and Pb has posed a potential carcinogenic threat to humans (Madhuri *et al.*, 2012c; Pandey *et al.*, 2013a and 2014). The presence of carcinogens, combined with fish tumour induction data and reports of tumours in wild fish populations have provided strong circumstantial evidence that chemical carcinogenesis is occurring at some sites. Common cancers reported in zebrafish are seminoma, HCC, HCA, adenoma of exocrine pancreas, intestinal adenocarcinoma, ultimobranchial neoplasm, thyroid neoplasm, spindle cell sarcoma and hemangioma. DEN is one of the most potent carcinogens. The most common type of interstitial cell neoplasm reported to occur in several fish species is hemangiopericytoma. Neoplasm has also been induced in intrahepatic and extrahepatic abdominal locations by MAMA and N-methyl-N'-nitro-N-nitrosoguanidine (MNNG). Epizootics of skin cancer have been associated with chemical contamination. Although environmental pollutants are important factors in the development of hepatic and skin cancers in fish, the causes may be multifactorial (Pandey *et al.*, 2013a, 2013b and 2014).

Chapter 2

Fish Research Model in Pharmacology and Toxicology

Fish Models in Pharmacological and Toxicological Researches

Fish as Experimental Model

Fish not only play an important role in the demand of food for humans but they are widely used as model organisms for various biological experiments. World of fish pharmacology is now changing quickly. Different experiments and the use of drugs in fish have been elucidated in various literatures. Recognizing differences in the physiology of fish is essential to understanding the problems associated with drug approval for the aquatic environment. Ectothermic animals change body temperatures through a wide range, affecting both the uptake and distribution and biotransformation of drugs. Antibacterial, antiparasitic and anaesthetic drugs, besides the pharmacokinetic and pharmacodynamic parameters have been well experimented on the fish. Several groups of drugs, *e.g.*, tetracyclines, penicillins, macrolides, quinolones, sulfonamides, immunostimulants, anticancer agents, herbal drugs and vaccines, have been successfully experimented or used in different species of fish. It would have been useful to do more pharmacological and toxicological experiments in fish on emerging new drugs. It will be nice to observe the effects of breeding induction agents, immunostimulants, anticancer agents, herbal drugs and vaccines, all used in fish husbandry and therapeutic strategies in aquaculture (Madhuri and Pandey, 2014; Pandey, 2011a).

In pharmacological and toxicological research, fish has also been used as a new model organism to create experimental carcinogenesis/cancer and ultimately to evaluate the anticancer activity of drugs. Potential for application of research findings to both human and environmental health issues makes fish species attractive and valuable alternative models in the carcinogenesis and toxicity research. Zebrafish (*Danio rerio;* Syn. *Brachydanio rerio*), a freshwater tropical fish, can be kept easily and cheaply in large numbers in laboratory, where it breeds all year round. Hence, zebrafish is recognized as a suitable model for different experimental studies (Bunton, 1996).

Fish is now used as the model organism for different experimental studies, including those of pharmacology and toxicology. Mangrove-dwelling fish, *R. marmoratus* (rivulus), is the only vertebrate that is a synchronous, internally self-fertilizing hermaphrodite. This unique reproductive mode yields offspring with little genetic variation, which offers significant advantage for the use of this species in bioassays. In the 1960s to mid-1970s, the fish species used as models for cancer study were primarily the zebrafish and guppy. Species that have predominated in later years can be divided into two groups: larger fish, *e.g.,* rainbow trout, and small aquarium fish, including rivulus, guppy, sheephead minnow and medaka. Contamination associated neoplasia, including aflatoxin-induced HCC in rainbow trout, has also led to the study of fish as alternative model in carcinogenesis and toxicity bioassay (Bunton, 1996; Pandey, 2011a).

Zebrafish has been reported to be an important model organism in the developmental genetics, neurophysiology and biomedicine (Madhuri and Pandey, 2014; Pandey, 2011a). Over 400 labs worldwide now routinely use the zebrafish in several researches, and there is an increasing interest in its use as a model for understanding the genetic basis of behaviour (Miklosi and Andrew, 2006). The zebrafish is increasingly important in biomedical research, particularly as a model of human disease and for the screening of therapeutic drugs (Rubinstein, 2006). Hence, zebrafish has emerged as a major model organism for biomedical research, and it is a premiere model organism to study the vertebrate development (Pandey, 2011a). Zebrafish may also be a powerful model for the study of human diseases because many cellular processes are conserved throughout vertebrate evolution, including corresponding disease genes (Willemsen *et al.,* 2011).

Fish as a Model Organism in Cancer Research

Fish have gained increasing attention over the past three decades as valuable models for environmental cancer/carcinogenesis research. Different fish species have been investigated as non-mammalian vertebrate models for carcinogen testing, as surrogates for understanding mechanisms of human cancer and its prevention, as feral species indicators of ecologic contamination, as indicators of potential human exposure to carcinogens in the water column or aquatic food chain, and for application as *in situ* field monitors of integrated carcinogenic hazard in ground water near toxic waste sites. Compared to mammals, cancer induction in fish appears to be easy. The carcinogen can be exposed through aquarium water and liver tumour induction, in particular, provides the best results with most of the carcinogens so far studied (Pandey *et al.,* 2013a; Raisuddin and Lee, 2008).

In the 1960s to mid-1970s, the fish species used as models for carcinogenesis studies were mainly the *D. rerio* and *P. reticulata*. Species that have predominated in later years are: the larger fish, *e.g.,* rainbow trout, and the small aquarium fish, including *R. marmoratus, P. reticulata, C. variegatus, Xiphophorus/Platypoecilus* species, *F. grandis, G. affinis, C. variegates* and *O. latipes*. Contaminations associated neoplasia, including aflatoxin-induced HCC in *O. mykiss* fish, have also led to the study of fish as alternative model organisms in cancer and toxicity bioassays (Madhuri and Pandey, 2012 and 2014; Pandey *et al.,* 2012f and 2013a).

Tumours in zebrafish can be generated by treatment with chemical carcinogens or by genetic approaches. Liver has been the chief target organ for tumorigenesis after carcinogen treatment while many other tissue-specific tumours have been generated by tissue-specific expression of proven oncogenes. A remarkable similarity has been demonstrated in the molecular hallmarks during liver tumorigenesis between humans and zebrafish, thus validating the zebrafish model for human cancer studies. Zebrafish models for liver cancers have been successfully established which are now increasingly used as a promising animal model for cancer research. Hence, these models will be characterized in order to understand the molecular and genetic mechanisms of liver carcinogenesis as well as for anticancer drug discovery (Gong *et al.,* 2011).

Cell lines provide an important biological tool for carrying out investigations into physiology, virology, toxicology, carcinogenesis and transgenics. Teleost fish cell lines have been developed from a broad range of tissues like ovary, fin, swim bladder, heart, spleen, liver, eye muscle, vertebrae, brain and skin. Over 124 new fish cell lines from different fish species ranging from grouper to eel have been reported. Presently, about 283 cell lines have been established from finfish around the world (Lakra *et al.,* 2011; Madhuri and Pandey, 2012 and 2014).

Fish Species in Cancer and Other Biological Researches

Different fish species have been recognized as model organisms for cancer and other biological researches. Compared to mammals, cancer induction in fish appears to be easy. The main fish species that have been used as models for cancer and other biological researches are zebrafish (*Danio rerio*- Figure 19), rainbow trout (*Onchorynchus mykiss*- Figure 20), mangrove rivulus or mangrove killifish (*Kryptolebias marmoratus,* formerly *Rivulus marmoratus*- Figure 21), medaka (*Oryzias latipes*- Figure 22), platyfish (*Xiphophorus maculates*), northern pike (*Esox lucius*), muskellunge (*Esox masquinongy*), goldfish (*Carassius auratus*- Figure 23), carp (*Cyprinus carpio*- Figure 24), walleye (*Sander vitreus,* formerly *Stizostedion vitreum*), winter flounder, black back or righteye flounder (*Pseudopleuronectes americanus*), Hawaiian pebbled or multiband butterfly fish (*Chaetodon multicinctus* and *C. miliaris*), guppy (*Poecilia reticulata*- Figure 25), sheephead minnow (*Cyprinodon variegatus*- Figure 26), gulf killifish (*Fundulus grandis*), mosquitofish or gambezi (*Gambusia affinis*), bicolor damselfish (*Pomacentrus partitus*) and nibe croaker (*Nibea mitsukurii*). One hundred twenty-four new fish cell lines have been reported, and about 283 cell lines have been established from finfish all over the world (Madhuri and Pandey, 2012 and 2014; Madhuri *et al.,* 2012h).

Some other fish species in which various biological experiments have been done are white sucker (*Catostomus commersoni*, a teleost fish- Figure 27), spotted snakehead (*Channa punctatus*- Figure 28), snakehead murrel, common snakehead, chevron snakehead or striped snakehead (*Channa striata*- Figure 29), rohu or roho labeo (*Labeo*

Figure 19: *Danio rerio.*

Figure 20: *Oncorhynchus mykiss.*

Figure 21: *Kryptolebias marmoratus* .

Figure 22: *Oryzias latipes.*

Figure 23: *Carassius auratus.*

Figure 24: *Cyprinus carpio.*

Figure 25: *Poecilia reticulata.*

Figure 26: *Cyprinodon variegates.*

Figure 27: *Catostomus commersoni.*

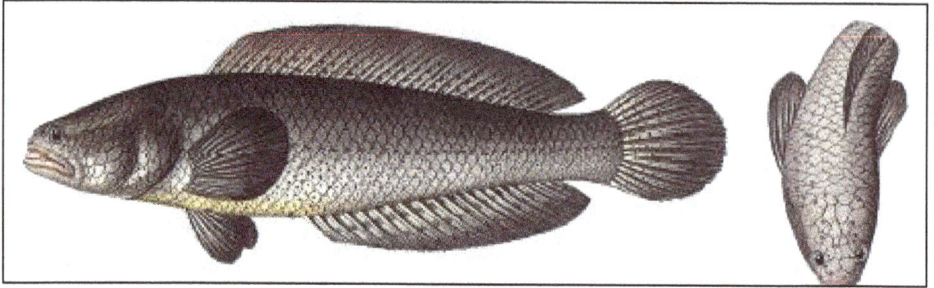

Figure 28: *Channa punctatus.*

Figure 29: *Channa striata.*

Figure 30: *Labeo rohita.*

Figure 31: *Oreochromis niloticus.*

Figure 32: *Fundulus heteroclitus.*

rohita- Figure 30), nile tilapia (*Oreochromis niloticus*- Figure 31), mozambique tilapia (*Oreochromis mossambicus*), mummichog (*Fundulus heteroclitus*- Figure 32), catla or katla (*Catla catla* or *Gibelion catla*- Figure 33), African sharptooth catfish (*Clarias gariepinus*), mrigal carp, white carp or mrigal (*Cirrhinus mrigala* or *C. cirrhosus*, an Indian major carp- Figure 34), red seabream (*Pagrus major*), pez chato, flat fish or pirai (*Piaractus mesopotamicus*), coho salmon (*Oncorhynchus kisutch*- Figure 35) and Guinean tilapia (*Tilapia guineensis*- Figure 36), etc.

Figure 33: *Catla catla.*

Figure 34: *Cirrhinus mrigala.*

Figure 35: *Oncorhynchus kisutch.*

Figure 36: *Tilapia guineensis.*

Here are some important fish species on which detailed cancer/carcinogenesis research was performed by various investigators:

1. *Danio rerio* (Zebrafish)

This freshwater tropical fish is a small, robust fish. This can be kept easily and cheaply in large numbers in the laboratory. Its generation time is short, typically 3 to 4 months, making it suitable for experiment. Mean life span of domesticated zebrafish has been reported to be 42 months, with the oldest fish surviving for 66 months. Zebrafish are omnivorous; and are floodplain species, abundant in shallow ponds, ditches or silt-bottomed, well-vegetated pools and rice paddies adjacent to slow moving streams at a range of elevations (Gerhard *et al.*, 2002). Zebrafish has been increasingly recognized as a promising animal model for cancer research. Tumours in zebrafish can be generated by treatment with chemical carcinogens or by genetic approaches. Liver has been a main target organ for tumorigenesis after carcinogen treatment. Thus, zebrafish model has been used for human cancer studies, particularly for liver cancer (*e.g.*, HCC) which can be used for anticancer drug discovery (Gong *et al.*, 2011; Madhuri, 2012; Pandey *et al.*, 2012f).

2. *Oncorhynchus mykiss* (Rainbow Trout)

This fish has been used as a model for the study of carcinogenicity caused by many food and environmental contaminants. This is a suitable alternative model for environmental carcinogenesis. Trout has a complex but incompletely characterized array of cytochrome P450, transferases and other enzymic systems for procarcinogen metabolism. In general, trout exhibits only limited capacity for DNA repair, especially for removal of bulky DNA adducts. This factor, together with a high capacity for P450 bioactivation and negligible glutathione transferase-mediated detoxication of the epoxide, accounts for the exceptional sensitivity of trout to aflatoxin B_1 (a toxin produced by fungus *Aspergillus flavus* on foods and feeds) carcinogenesis. At the gene level, all trout tumours except nephroblastoma exhibit variable and often high incidences of oncogenic Ki-ras gene mutations. There are many aspects of the trout model, especially the lack of complete organ homology, that limit its application as a surrogate for human cancer research. Within these limitations, however, it is apparent that trout and other fish models can serve as highly useful adjuncts to conventional rodent models in the study of environmental carcinogenesis and its modulation. In the embryonic exposure to aflatoxin B_1, the rainbow trout produced a high incidence of HCC in adults (Madhuri and Pandey, 2012). Rainbow trout also pioneered the research mostly on liver cancer development and prevention. Toxicogenomics study of transcriptional patterns in aflatoxin B_1-induced HCC in rainbow trout and human HCC showed a high degree of similarity (Tilton *et al.*, 2005).

3. *Kryptolebias marmoratus* (Mangrove Rivulus or Mangrove Killifish)

This is the only known vertebrate with internal self-fertilization. Because of this peculiar reproduction and distinctive habitat, *K. marmoratus* has attracted interest from some research groups for its use in toxicology and carcinogenesis research. Laboratory maintenance of *K. marmoratus* is easy, and it is a kind of euryhaline fish able to thrive in a broad salinity range. Exposure to endocrine disrupting chemicals in environmental carcinogenesis shows that *K. marmoratus* modulates the expression of genes critical in tumour development and regulation. Development of chemically-induced tumours and its transplantation as shown in case of thyroid tumour are important features of *K. marmoratus* for its consideration and development as model species for cancer research. Recent studies on sequence and expression data of oncogenes (N-*ras* and R-*ras*, in particular) and tumour suppressor gene p53 have further enriched our knowledge about oncogenic responses in *K. marmoratus*. All these features make the use of *K. marmoratus* in experimental cancer research a good choice. These studies highlight that *K. marmoratus* shows a good tumour induction response even when exposed to a single carcinogen dose (Raisuddin and Lee, 2008). Usefulness of *K. marmoratus* in cancer research for the first time was highlighted in 1984 (Koenig and Chasar, 1984).

4. *Oryzias latipes* (Medaka)

It is a small freshwater aquarium fish, native to Japan and Asia. It is highly useful in cancer research. Utility of medaka for carcinogenesis research was initially demonstrated in 1975 (Ishikawa *et al.*, 1975), and since then medaka have been proved to be a sensitive subject for a variety of carcinogens in an equally varied number of organ systems (Okihiro and Hinton, 1999). Medaka has emerged as the small fish of choice in many carcinogenesis bioassays. A relatively short period of time, short reproductive cycle, hardiness under test conditions and easy maintenance are also the attractive characteristics of medaka. These factors can decrease the overall costs of performing assays. Major disadvantage, however, is also related to size, which so far has somewhat limited the breadth of studies performed on this and other aquarium species (Bunton, 1996).

5. *Xiphophorus maculatus* (Platyfish)

It is one of the oldest and best groups of inbred strains. This fish is small (typically <50 mm standard length), and is derived from freshwater habitats in Mexico, Guatemala, Belize and Honduras. Use of these fishes as a research model to study the genetic components of carcinogenesis has a history encompassing more than 70 years. In the 1920s and early 1930s, the American biologist Myron Gordon and German scientists G. Haussler and C. Kosswig independently discovered that interspecies hybrids between strains of the southern platyfish *X. maculatus* and the green swordtail *X. helleri* develop melanomas spontaneously. These cancers develop after interspecific crossing and originate from extreme phenotypic enhancement of melanistic '*macromelanophore*' pigment derived from southern platyfish. Melanomas, which arise from phenotypic over expression of macromelanophore pigment and interspecific crossing, typically exhibit a disproportionate number of melanocytes which actively proliferate without sufficient regulation (Bunton, 1996; Madhuri and Pandey, 2012).

6. Other Fish Species

Majority of cancers in fish associated with environmental chemical contamination are of epithelial origin, with liver and skin as the primary affected organs. Many epizootics of non-epithelial cancers like lymphoma in northern pike (*E. lucius*) and muskellunge (*E. masquinongy*), neurofibroma and neurilemmoma in goldfish (*C. auratus*) and bicolor damselfish (*P. partitus*), gonadal neoplasms in carp (*C. carpio*), and dermal fibroma and fibrosarcoma in walleye (*S. vitreus*) have a stronger association with viral aetiology or genetic predisposition. Epizootic hepatic neoplasia was also reported from the North Atlantic coast in winter flounder (*P. americanus*), a commercially valuable, benthic (bottom dwelling) fish. A necrogenic phase was also seen with other chemicals like MAMA, and was described in English sole, winter flounder and hatchery rainbow trout from neoplasia epizootics. Some pigment cell neoplasms, including epizootic chromatophoroma in the nibe croaker (*N. mitsukurii*) and Hawaiian pebbled or multiband butterfly fish (*C. multicinctus* and *C. miliaris*) were related to sediment contaminated by pulp mill effluent and agricultural runoff, respectively (Bunton, 1996; Madhuri and Pandey, 2012).

Some Pharmacological and Toxicological Studies in Fish

Innervations of the heart of trout (*Salmo trutta* and *S. iredius*) have been studied with fluorescent histochemical and physiological methods. Catecholamine-containing nerves were revealed by the histochemical technique in the sinus venosus, atrium and ventricle. Stimulation of the vagus nerve or the ductus Cuvieri (especially at high-pulse frequency and duration) produced inhibition of the heart, which was blocked by atropine. After atropine treatment, an excitatory response was revealed which was blocked by guanethidine, bretylium or pronethalol. Finally, it has been suggested that the cardiac branch of the vagus nerve contains both inhibitory cholinergic and excitatory adrenergic nerves. The release of endogenous dopamine from teleost retina was studied using high-performance-liquid-chromatography and electrochemical detection. The light evoked release of dopamine was inhibited by GABA and l-glutamate, whilst antagonists of these retinal transmitters stimulated the release in dark. Aminergic metabolism was studied in discrete brain regions of post-ovulated female rainbow trout using a liquid chromatography electrochemical detection method. 3-methoxytyramine (3-MT) was the major dopaminergic catabolite, suggesting that catechol-o-methyl transferase (COMT) is the main dopamine (DA) catabolic enzyme. Two populations of brain regions were found: one with a high DA content and low 3-MT/DA ratio (hypothalamus and telencephalon), suggesting that these regions could present a high density of DA perikarya; the other with a high 3-MT/DA ratio (pituitary, preoptic area, myelencephalon and optic tectum), suggesting that these regions could present a high density of DA axonal endings. 5-hydroxytryptamine (5-HT) content differed, but a homogeneous distribution of monoamine oxidase was seen in different regions. High 5-HT content was found in hypothalamus and telencephalon (Madhuri and Pandey, 2014; Pandey, 2011a).

Effect of salinity on the acute toxicity of Cd was studied in the tropical, estuarine, hermaphroditic fish, *R. marmoratus*. Live fibre bundles were isolated from fast myotomal muscle of short-horned sculpin (*Myoxocephalus scorpius*) marine fish, and isometric contractile properties and force-velocity relationship were determined at 5, 10 and 15°C. Maximum contraction speed of muscle fibres at 15°C was 2.4 times higher in 15°C- than in 5°C- acclimated fish. So, acclimation modified the contractile properties of fast muscle fibres both at low and high temperatures. Another important experiment done in fish was the study of molecular evolution of opioid receptor family. cDNAs that encode 6 distinct opioid receptor-like proteins were isolated from *C. commersoni* teleost fish. One of these encodes a 383-amino acid protein that shows greatest sequence similarity to mammalian opioid (m-opioid) receptors; corresponding gene is expressed predominantly in brain and pituitary. Transfection of teleost cDNA into HEK293 cells resulted in the appearance of a receptor with high affinity for m-selective agonist and non-selective antagonist naloxone. Data also indicated that m-opioid receptor arose very early in evolution, perhaps before appearance of vertebrates, and that pharmacological and functional properties of this receptor have been conserved over a period of 400 million years implying that it fulfils an important physiological role. Various drugs were evaluated against *H. salmonis* infection in rainbow trout. Results confirmed the efficacy of nitroimidazoles, and infection was completely eradicated not only by metronidazole (which was recommended for hexamtosis treatment), but also by benznidazole, ronidazole and secnidazole. Non-nitroimidazoles albendazole, amnosidine, hethylcarbamazine and nitroscanate also completely eliminated the infection (Pandey, 2011a).

Venomous creatures have been the source of much recent research in the effort to find novel physiological tools and pharmaceuticals. Thus, an experiment was successfully performed in fish (stonefish), suggesting that both functionally in experimental models and in western immunoblotting analysis, piscine venoms may possess structural as well as functional similarities as compared to the venoms of other species. The most potent effects of piscine venoms are on the cardiovascular system (Church and Hodgson, 2002).

An experiment on fish was conducted for pharmacological characterization of melanocortin (MC) receptors suggesting the role for ACTH. The MC receptor subtypes have distinctive characteristic binding profiles. The MC4 receptors of trout and fugu fishes had similar affinity for α-MSH and β-MSH, and a much higher affinity for ACTH than does the human MC4 receptor. The fugu MC1, and the trout and fugu MC5 receptors also had higher affinity for ACTH-derived peptides than α-, β-, or γ-MSH. The ACTH-derived peptides might have played an important role at the MC receptors, while the specificity of the different subtypes for α-, β-, and γ-MSH peptides might have appeared at later stages during vertebrate evolution (Haitina *et al.*, 2005). On the premise that human medicines may potentially induce similar pharmacological and toxicological profiles in fish and other lower vertebrates, a comparative approach was applied to β-adrenergic receptor antagonists (β-blockers) which are widely detected in surface water. Activities of these compounds may be observed more efficiently in fish than in mammals. Extensive mammalian pharmacological and toxicological studies are central to development of medicines and these can provide valuable information to guide ecotoxicological studies. Some β-adrenergic receptors have been characterized in fish using both traditional molecular cloning methods, or via mining of genomic sequences from various organisms. These approaches elicit that fish have β-adrenergic receptors very similar to those present in mammals. Since any effects of β-blockers in fish are most likely to be mediated via β-adrenergic receptors, it is the physiological processes regulated by these receptors that are most likely to be affected. Thus, cardiovascular dysfunction is one possible consequence of exposure of fish to these compounds, leading to impaired fitness (*e.g.*, reduced growth and fecundity) (Owena *et al.*, 2007). In a study on the spotted snakehead (*C. punctatus*) fishes, it was seen that when these fishes were exposed to high concentration (2 mM) of sodium arsenite, they died within 2.5 hr. Chromosomal DNA of liver cells were fragmented which suggest that sodium arsenite might induce death of those cells through apoptosis (Ahmed *et al.*, 2008). Toxicity of sodium cyanide (NaCN, free cyanide) to freshwater fish, *L. rohita* was studied and the LC_{50} of NaCN in 96 hr was evaluated as 33 μg/L. Behavioural changes in fishes were also noticed after the lethal and sublethal doses of NaCN (Prashanth *et al.*, 2011).

Experimental studies have shown the induction of cancer in fish by chemical carcinogens. DEN, an N-nitroso (nitrosamine) compound, is one of the most potent carcinogens. Primary neoplasms with histological characteristics of HCC were observed in fish exposed to 125 ppm of DEN for 3 to 5 successive periods. After administration of DEN (95 mg/L for 6 weeks), hemangiomas, cholangiomas, biliary cystadenomas and HCC were seen in the liver tissues of fish at the 18th week. Rainbow trout were fed with a diet containing indole-3-carbinol (2000 ppm), β-naphthoflavone (500 ppm) or Aroclor 1254 (100 ppm) for 6 weeks before a single 24 hr exposure of an aqueous solution of 250 ppm DEN. After 42 weeks, DEN produced 80.2 per cent incidence of liver tumours. The tumour was inhibited by indole-3-carbinol but enhanced by β-naphthoflavone. Liver cytotoxic alterations of adult medaka fish were seen following short-term bath exposure (48 hr) to 500 mg/L DEN for 3 to 21 days. Hepatomas were noticed in 6- and 9- month groups of the 100 ppm- and 200 ppm-DEN treated fishes. Guppies were exposed to multiple doses of DEN. Neoplastic foci of mixed hepatocytes and cholangiocytes increased in livers of guppies from the 2nd month, developing into hepatoblastomas, which occurred in almost 100 per cent of guppies by the 12th month. HCC was seen in rainbow trout fish exposed to the organochlorine insecticide, DDT. Neoplasms in rainbow trout exposed to MNNG were seen in liver, stomach and kidney, with a low incidence in swim bladder. Methylazoxymethanol acetate (MAM-Ac), a potent carcinogen, also caused cancer in fish, but to lesser degree than with MNNG. Liver was most commonly affected, followed by muscle (smooth and skeletal muscle, including heart), pancreas, connective tissues and kidney. Indirect-acting carcinogens like PAHs, with B(a)P and 7,12-dimethylbenzanthracene (DMBA) have also been used to cause fish cancer. Rivulus fish was also used to induce the HCC by dietary antioxidant butylated hydroxyanisole (Bunton, 1996; Pandey, 2011a). Progression of hepatic neoplasia was observed in adult medaka fish (3-6 months old) following aqueous exposure to DEN (50 ppm for 5 weeks) (Okihiro and Hinton, 1999). DEN-induced spongiosis hepatis (a hepatic lesion characterized by mutilocular cyst-like complexes) was noticed in the Japanese medaka fish (Norton and Gardner, 2005).

Medicinal plants are alternative to antibiotics for fish health management. They do not only safe to consumers, but also have significant role in aquaculture. Many studies have been performed in different fish species. Acute toxicity of Indian almond and garlic was tested in tilapia fingerlings. It was shown that these herbs have low acute toxicity and can treat the trichodiniasis caused by *Trichodina*. The immunostimulant effects of the dietary intake of *Viscum album, Urtica dioica* and *Zingiber officinale*) on rainbow trout have also been studied (Chitmanat *et al.,* 2005). Immunostimulatory effect of aqueous extract of *Eclipta alba* leaf in tilapia fish, *O. mossambicus* was observed (Christybapita *et al.,* 2007). The antibacterial activities of the methanolic extracts of 31 Brazilian plants against fish pathogenic bacteria have been screened out (Castro *et al.,* 2008).

Immunostimulatory effects of two Chinese herbs (*viz., Lonicera japonica* and *Ganoderma lucidum*) were determined in tilapia fish (*O. niloticus*). It was observed that oral administration of ginger extract increases the phagocytic capability of cells in rainbow trout, while the extracts of four Chinese herbs (*Rheum officinale, Andrographis paniculata, Isatis indigotica* and *L. japonica*) increased the phagocytosis of white blood cells (WBCs) of carp (Yin *et al.,* 2008). Goldfish was used to observe the effects of *Phyllanthus niruri* and *Aloe vera*, and it was found that these herbs can positively enhance the growth performance of fish, as well as they act against *A. hydrophila* infections. Synergistic effect of herbs has also been noted in other fish species, including Japanese flounder and *Clarias gariepinus*. Growth increase in *L. rohita* fish fed with herbal supplemented diet was also recorded. An experiment was also conducted on the *Catla catla* fish, and the disease resistant was produced through immersion treatment of three herbs, *viz., A. sativum, A. indica* and *Curcuma longa* (Haldi) in spawn (Ahilan *et al.,* 2010). Fingerlings of Indian major carp (*L. rohita*) were experimented for the immunostimulatory and disease resistance effects of *Withania somnifera* (Ashwagandha) root against *A. hydrophila* infection. *L. rohita* Fingerlings were also used to evaluate the immunostimulant effects of the dietary intake of *A. sativum* and *Vitex negundo* extracts (Pandey *et al.,* 2012b; Sharma *et al.,* 2010). Pharmacologically, many plant products have been proved to be the potent antiviral drugs against fish viruses (Pandey, 2011a).

Zebrafish Model in Biomedical Research

A Model Organism

'*A model organism*' is the organism with a wealth of biological data that makes it attractive to study as an example for other species and/or a natural phenomenon that is more difficult to study directly. Historically, the model organism includes a handful of species with extensive genomic research data. Hence, a model organism is a non-human species that is studied to understand a particular biological phenomenon with the expectation that discoveries made in the organism model will provide insight into the workings of other organisms (Fields and Johnston, 2005). Model organisms are widely used to explore potential causes and treatments for human disease when the human experimentation would be unfeasible or considered less ethical (Fox, 1986).

Zebrafish

Zebrafish (zebra fish or zebra danio, *Danio rerio*) is a very popular freshwater tropical fish. Therefore, zebrafish is a popular aquarium fish, frequently sold under the trade name *'zebra danio'*. It is a small, robust fish, and so large numbers can be kept easily and cheaply in laboratory, where it breeds all year round (Pandey, 2011a and 2011b). It has a nearly transparent body during early development, which provides unique visual access to the animal's internal anatomy. Humans and zebrafish have shared waterways around the river Ganges for tens of thousands of years (Pandey *et al.*, 2012f).

There are a long row of different variants of this fish available in addition to the regular form. These forms include classical forms like long finned zebrafish, albino zebrafish, golden zebrafish, sandy zebrafish and leopard danios (earlier known as *Danio frankeri*), but also modern genetically engineered zebrafish that glows in the dark. The glowing variant is available in red, yellow and green and is often sold under the name *'glowfish'* (the name is a trademark for these patented fish). Glowfish was not designed for the aquarium market but is proving popular and might lead to more design species being developed for the hobby in the future. Recently, transgenic zebrafish is commercially available that express green fluorescent protein, red fluorescent protein and yellow fluorescent protein. These transgenic zebrafish are trade named as *'glowfish'*. Other cultivated varieties of zebrafish are golden, sandy, longfin and leopard (Madhuri, 2012; Pandey *et al.*, 2012f). The leopard danio strain is a spotted colour morph of zebrafish caused by a pigment mutation. The xanthistic forms of both the zebra and leopard pattern, along with long-finned subspecies, have been obtained via selective breeding programmes for the aquarium trade. Zebrafish has its wild-type strains, the most common are AB, TU, IN and WIK strains. The hybrids between different *Danio* species may be fertile, *e.g.*, between *D. rerio* and *D. nigrofasciatus*. A zebrafish pigment mutant, called *'bleached blond'* was produced by insertional mutagenesis (Madhuri, 2012).

Zebrafish is a small, shoaling cyprinid, native to the flood-plains of the Indian subcontinent, where it is found in shallow, slow-flowing water. Zebrafish belongs to minnow family (Cyprinidae) of Order Cypriniformes (freshwater fishes), the most species of which are rich vertebrate family. The name *'Danio'* derives from the Bengali name *'dhani'*, meaning *'of the rice field'*. The two genera, *viz.*, *Devario* and *Danio* of zebrafish are ecologically quite distinct, *Devario* species occurring in hill streams with clear running water, while *Danio* species are confined to lowland areas, typically inhabiting slow-flowing, turbid rivers and pools. Other danionin species found with zebrafish are *Danio dangila*, *D. meghalayensis*, *Devario devario*, *De. assamensis* and

De. aequipinnatus. Zebrafish are characterized by small, slender shape (up to 4 cm long, or <120 mm total length), the presence of a *'danionin notch'* in the ventro-medial margin of dentary and a distinctive colour pattern based on alternating dark and light horizontal stripes, which may be broken up into blotches or bars. In other words, zebrafish have luminous bluish-black and silvery-gold horizontal stripes or dark-blue and silvery longitudinal stripes. It has nearly transparent body during early development, which provides unique visual access to animal's internal anatomy. Natural range of zebrafish is centred around the Ganges and Brahmaputra river basins in north-eastern India, although the specimens were also collected in the Indus, Cauvery, Pennar, Godavari and Mahanadi river basins. In addition, it lives in Krishna river basin. Conclusively, zebrafish originate from eastern Asia, and distributed throughout the south and south-east Asia, their highest species diversity in north-eastern India (including Rajasthan, Gujarat, Andhra Pradesh, West Bengal, Assam, Meghalaya and Orissa), Pakistan, Bangladesh, Nepal, Burma (Myanmar) and Sri Lanka. It is commonly found in most types of water ranging from streams to stagnant ponds and rice fields. It has also been accidentally introduced into a number of different waters around the world and it can today also be found in countries like USA, Australia and Colombia (Madhuri, 2012; Pandey *et al.*, 2012f).

Zebrafish is named for the five uniform, pigmented, horizontal blue stripes on the side of the body, all of which extend to the end of the caudal fin. Its shape is fusiform and laterally compressed, with its mouth directed upwards. Males are torpedo-shaped and have gold stripes between the blue stripes; females have a larger, whitish belly and have silver stripes instead of gold. Adult females will exhibit a small genital papilla in front of the anal fin origin. Life-span in captivity is around 2 to 3 years, although in ideal conditions, may extend to 5 years (Spence *et al.*, 2008). According to Gerhard *et al.* (2002), the mean life span of domesticated zebrafish is 42 months (3.5 years), with the oldest fish surviving for 66 months (5.5 years).

Further, zebrafish are found in rivers, small streams and other channels, stagnant or slow-moving pools near streams, and rice paddies (Daniels, 2002). Zebrafish appear to be primarily a floodplain species, inhabiting shallow ponds and ditches or slower reaches of streams. They are an abundant species and are among the smallest fish species in the assemblages in which they occur. They feed throughout the water column and tend to be confined to shallow margins of water bodies. They are particularly abundant in silt-bottomed, well-vegetated pools and rice paddies adjacent to slow moving streams at a range of elevations. They are commonly inhabited in streams, canals, ditches, ponds and slow-moving to stagnant water bodies, including rice fields (Spence *et al.*, 2008).

The breeding season of zebrafish is between April and August (Engeszer *et al.*, 2007), presumably varying somewhat by latitude, elevation and prevailing climatic conditions. It can grow to 6.4 cm (2.5 inches) (Spence *et al.*, 2008). Generation time of zebrafish is short, typically 3 to 4 months, making it suitable for experiment. A male must be present for ovulation and spawning to occur. Females are able to spawn at intervals of 2 to 3 days, laying hundreds of eggs in each clutch. Upon release, the embryonic development begins; absent sperm, growth stops after the first few cell divisions. The fertilized eggs almost immediately become transparent, a characteristic which makes zebrafish a convenient research model species. Its development is rapid, with precursors to all major organs developing within 36 hr of fertilization. Hatching takes place 48 to 72 hr after fertilization, depending on embryo's internal conditions and external temperature, ideally 28.5°C (83.3°F). The larvae display food seeking, swimming and other active avoidance behaviours within 5 days post-fertilization, *i.e.*, 2 to 3 days after hatching. The sex of juveniles cannot be distinguished except by dissection, and sex determinants are not clearly understood. Egg laying is thought to occur in small pools adjacent to streams. Growth rate of zebrafish is most rapid during first 3 months following hatching, after which it starts to decrease, reaching zero by about 18 months (Daniels, 2002; Spence *et al.*, 2008; Madhuri, 2012; Pandey *et al.*, 2012f).

Zebrafish have the ability to regenerate fins, skin, heart and brain (in larval stages). The heart muscle regeneration does not make use of stem cells; instead, mature heart muscle cells regress to a stem-cell-like state and re-differentiate. Zebrafish regenerate photoreceptor cells and retinal neurons following injury. Mechanisms of this regeneration are unknown. Researchers frequently amputate dorsal and ventral tail fins and analyze their re-growth to test for mutations. This research is leading the scientific community in understanding of healing/repair mechanisms in vertebrates. If the same fin is damaged enough times, the fish will grow a new fin which will have mutated only a small amount. Most scientists believe that this is a defence mechanism to try to prevent fin from being damaged again (Madhuri, 2012).

Zebrafish are omnivorous; their natural diet consists mainly of zooplankton and insects; though phytoplankton, filamentous algae and vascular plants, spores and invertebrate eggs, fish scales, arachnids, detritus, sand and mud have also been reported from gut content analysis. Zebrafish themselves are known to feed on mosquito larvae and, presumably, other insects, although the precise species are not known (Daniels, 2002; Pandey *et al.*, 2012f). According to Spence *et al.* (2007) also, zebrafish is omnivorous which primarily eats zooplankton, insects and insect larvae, and phytoplankton; and can also eat a variety of other foods like worms and small crustaceans if its preferred sources are not readily available. Most of them accept common food flakes and tubifex worms in the aquarium (Madhuri, 2012).

Zebrafish is popular in home aquaria because it is a friendly fish, and is very hardy and adaptable which makes it an ideal beginner fish. It is said to be a very popular fish as it can be found in just about any aquarium shop around the world. Zebrafish are breed in huge numbers for the aquarium market. Their ease of keeping and breeding, cold resistance, beauty, price, playful nature, and broad availability, all contribute to their popularity (Madhuri, 2012; Pandey *et al.*, 2012f). They thrive best in water at 22 to 28°C (72-82°F). They need an aquarium of 38 L (10 US gallons) or more, and they do well in schools. They also thrive as shoals of 6 or more and interact well with other fish types in the aquarium. However, they are susceptible to *Oodinium* or velvet disease, microsporidia (*Pseudoloma neurophilia*) and *Mycobacterium* species. Given the opportunity, adults eat hatchlings, which may be protected by separating the two groups with a net, breeding box or separate tank (Madhuri, 2012).

Zebrafish as Suitable Model Organism for Research

Zebrafish is known for its rapid development in the laboratory. The greatest advantage of zebrafish as a model system comes from its well-characterized genetics, genetic and developmental techniques and tools, and availability of well-characterized mutants. Like most other biomedical model organisms, zebrafish has been chosen for particular traits that make it convenient for lab study, not for a broad understanding of the organism in its native environment (Pandey *et al.*, 2013a; Webb and Schilling, 2006).

Zebrafish is used as a model organism in various biological studies, including study of specific gene function and roles of signaling pathways. Induction of experimental diseases, including cancer using zebrafish as a model organism may be a new and an important field of research. Zebrafish is an important model organism in developmental genetics, neurophysiology, oncology and biomedicine. This has emerged as an important vertebrate model organism in research that has allowed advances in the fields of toxicology, evolutionary theory, reproductive studies, teratology, genetics, neurobiology, environmental science, stem cell and regenerative medicine, and developmental biology and oncology. Thus, zebrafish are very popular among the scientists as they are a perfect model organism for vertebrates, and can be used as a complement to mice and rats in research. Zebrafish has proven very useful for research, and the study of this fish has formed the basis for many new promising medical technologies (Madhuri, 2012; Madhuri *et al.*, 2012h; Pandey, 2011b; Pandey *et al.*, 2012f and 2013a).

Potential for application of research findings to both human and environmental health issues make fish attractive and valuable model in the carcinogenesis and toxicity research. The strength of zebrafish as a model organism is that as a vertebrate, it is more comparable to humans than invertebrate model species such as (Barbazuk *et al.*, 2000), while being more tractable to genetic and embryological manipulation than mammalian model species such as mice, in which such procedures are both more complicated and costly. Zebrafish is increasingly important in biomedical research, particularly as a model of human disease and for the screening of therapeutic drugs (Rubinstein, 2006). Over 400 labs worldwide now routinely use zebrafish in several researches, and there is an increasing interest in its use as a model for understanding the genetic basis of behaviour (Miklosi and Andrew, 2006). Thus, zebrafish is a premiere model organism to study the vertebrate development. It may also be a powerful model to study the human diseases because many cellular processes are conserved throughout vertebrate evolution, including corresponding disease genes (Willemsen *et al.*, 2011).

The greatest advantages to use zebrafish as a research model are (Madhuri, 2012):

a) Zebrafish are fully-sequenced genomes.

b) They are well understood, easily observable and have testable developmental behaviours.

c) They have well-characterized mutants.

d) They have rapid embryonic development (progressing from eggs to larvae in 3 days, although overall generation time is comparable to that of mice).

e) They have large, robust and transparent embryos that develop outside the mother.

f) They have about constant size in early development that facilitates staining.

g) Drugs may be administered to zebrafish by adding directly in the tank.

h) Their unfertilized eggs can be made to divide.

i) Their two-celled embryo can be fused into a single cell, creating a homozygous embryo.

j) They have similarity to mammalian models and humans in toxicity testing.

k) They exhibit a diurnal sleep cycle similar to mammalian sleep behaviour and physiology.

In October 2001, researchers of the University of Oklahoma published *D. rerio's* mitochondrial DNA sequence (Broughton *et al.*, 2001). Its length is 16,596 base pairs. This is within 100 base pairs of other related fish species, and it is only 18 pairs longer than the goldfish (*C. auratus*) and 21 longer than the carp (*C. carpio*). Its gene order and content is identical to common vetebrate form of mitochondrial DNA. It contains 13 protein-coding genes and a non-coding control region containing origin of replication for heavy strand. In between a grouping of 5 tRNA genes, a sequence resembling vertebrate origin of light strand replication is found. It is difficult to draw evolutionary conclusion because it is difficult to determine whether base pair changes have adaptive significance via comparisons with other vertebrates nucleotide sequences (Madhuri, 2012). In December 2005, a study of golden strain identified the gene responsible for its unusual pigmentation as SLC24A5, a solute carrier which appeared to be required for melanin production, and confirmed its function with a Morpholino knockdown. Orthologous gene was then characterized in humans, and a one base pair difference was found to strongly segregate fair-skinned Europeans and dark-skinned Africans (Lamason *et al.*, 2005).

Transgenesis is a popular approach to study the function of genes in zebrafish. In January 2007, Chinese researchers at Fudan University genetically modified fish to detect oestrogen pollution in lakes and rivers, which is linked to male infertility. In 2008, researchers developed a new strain of zebrafish, named Casper, whose adult bodies were transparent. This allows for detailed visualization of cellular activity, circulation, metastasis and many other phenomena. Because many gene functions are shared between fish and humans, Casper is expected to yield insight into human diseases like leukemia and other cancers (Madhuri, 2012). Further in 2007, researchers at University College of London grew a type of zebrafish adult stem cell found in the eyes of fish and mammals that develops into neurons in the retina- the part of the eye which sends messages to the brain. These cells could be injected in the eye to treat diseases which damage retinal neurons- nearly every disease of the eye, including muscular degeneration, glaucoma and diabetes-related blindness. Retinal damage is responsible for most cases of sight loss. Researchers studied Muller glial cells in the eyes of humans aged from 18 months to 91 years, and were able to develop them into all types of retinal neurons. They were able to grow them easily in the lab. The stem cells successfully migrated into diseased rats' retinas and took on the characteristics of surrounding neurons (White *et al.*, 2008).

In cardiovascular research, zebrafish is being used to model blood clotting, blood vessel development, heart failure and congenital heart disease. The researchers have established a zebrafish model of inflammation and its resolution. In probing disorders of the nervous system, including neurodegenerative diseases, movement disorders, psychiatric disorders and deafness, researchers are using zebrafish to understand how the genetic defects underlying these conditions cause functional abnormalities in the human brain, spinal cord and sensory organs. The scientists are delving into the complexities of muscle degeneration in genetic models of human musculo-skeletal diseases, *e.g.*, muscular dystrophy. These studies exemplify how zebrafish research offers unique opportunities to understanding some of the greatest challenges in medical science, by enabling new discoveries of how cells and tissues work to be translated into new ways of understanding a broad range of human diseases. Finally, zebrafish models afford an ideal opportunity, not only to identify novel candidates for genes underlying human disease, but offers a potential system in which to begin to develop novel therapeutic agents in drug discovery programmes, hence helping to identify new treatments (Madhuri, 2012).

Last several years have witnessed an explosion in our understanding of vertebrate development, largely based on work from a few model genetic organisms. The embryonic zebrafish is one of these new models, and many research grants are being funded to utilize this model. Studies of the embryology and genetics of zebrafish, and identification of over 4,000 genetic mutations have lead to this dramatic increase in the number of laboratories (now about 250) using zebrafish to study the basic mechanisms of vertebrate development. Although zebrafish have become an extremely important research model, relatively little is known about diseases that affect this species when held in captivity. Indeed, severe health problems in zebrafish research colonies could seriously jeopardize many millions of dollars, and some researchers have experienced devastating, acute mortalities in their zebrafish colonies. Moreover, persistent, but less severe, infections have plagued several facilities. As with other laboratory animals used in research, it is imperative to conduct studies with disease-free, healthy zebrafish (Madhuri *et al.*, 2012h).

Zebrafish in Cancer Research

Experimental carcinogenesis using fish species as alternative models may be a new and an important field of research. Expansion of synthetic chemical producing industries during the 1940's coincided with a number of pollution-associated fish neoplasia epizootics, with PAHs as significant components of contaminated sediments in several cases. Epizootics of primarily liver and skin neoplasia in benthic species near coastal urban or industrial areas indicated the sensitivity of fish species to known mammalian carcinogens (Bunton, 1996).

Zebrafish has been increasingly recognized as a promising animal model for cancer research. Tumours (or cancers) in zebrafish can be generated by treatment with chemical carcinogens or by genetic approaches. Liver has been a main target organ for tumorigenesis after carcinogen treatment; while many other tissue-specific tumours have been generated by tissue-specific expression of proven oncogenes. Both chemical and transgenic approaches were used to generate liver tumours. By comparative analyses of transcriptome profiles between human liver tumours and carcinogen-induced zebrafish liver tumours, a remarkable similarity has been demonstrated in the molecular hallmarks during human liver tumorigenesis and carcinogen-induced zebrafish liver tumour, thus validating the zebrafish model for human cancer studies. The transgenic zebrafish models for liver cancers, including HCC have been established which will be further used to understand the molecular and genetic mechanisms of liver carcinogenesis as well as for anticancer drug discovery (Gong *et al.*, 2011; Madhuri, 2012; Madhuri *et al.*, 2012h; Pandey, 2011a). Zebrafish has been used to understand how a gene called *'Hedgehog'* (a biological signal that underlies a number of human cancers) controls the cell growth. This is fast becoming an avenue of research that is leading to new cancer therapies (Madhuri, 2012). Therefore, the zebrafish has been recognized as a suitable model for different experimental studies (Pandey, 2011a and 2011b).

Pioneering work by Dr. Mearle Stanton in the 1960s was the first research demonstrating experimental carcinogenesis in fish. This work showed that the zebrafish developed liver neoplasm (tumour) after exposure to DEN or methylazoxymethanol acetate (Stanton, 1966). Later, Drs. Jan Spitsbergen and Jerry Hendricks of Oregon State University showed that zebrafish of a Florida wild-type line (from a commercial wholesaler) are quite responsive to most carcinogens if exposed early in life. Liver was the most commonly targeted organ in these carcinogen studies with zebrafish, but as with other carcinogenesis studies in small aquarium fish, a wide variety of epithelial, mesenchymal, neural and neural crest neoplasms (tumours) occurred in fish exposed to many of the carcinogens. A study at Pennsylvania State University found a 100 per cent incidence of cutaneous papillomas in 18 zebrafish by one year after three immersion exposures at the adult stage to 2.5 to 3 mM ethylnitrosourea. The strain of zebrafish used in this papilloma study was not specified, but was presumably Florida wild-type as they were obtained from North American wholesaler. At Oregon State University, the spontaneous rate of cancer in the Florida wild-type line fed a semi-purified diet and held in a flow-through system was approximately 1 per cent at 6 to 14 months of age based on 3,000 untreated control fish. The most common spontaneous neoplasms at 6 to 14 months of age were seminoma, HCA and adenoma of exocrine pancreas, with intestinal adenocarcinoma being less common. A study of 21 Florida wild-type zebrafish over 18 months of age by Dr. Gary Marty indicated that 44 per cent of males had marked hyperplasia or neoplasm of seminiferous epithelium of testis. The most common target tissues for spontaneous neoplasia are testis, gut, thyroid, liver, peripheral nerve, connective tissue and ultimobranchial gland. Less common target tissues include blood vessels, brain, gill, nasal epithelium and the lymphomyeloid system. Although the investigators are just beginning to accumulate data regarding age-specific incidences of neoplasms in various lines of zebrafish, it is clear that the incidence of neoplasm or cancer increases strongly with age over one year in the lines examined. For example, the incidence of seminomas (neoplasms of the seminiferous epithelium of testis) is typically less than 1 per cent in fish of all lines examined so far at less than one year of age. But the incidence of seminomas increases to approximately 40 per cent by two years of age or more (Madhuri *et al.*, 2012h).

Aetiology of spontaneous neoplasms in zebrafish is unknown. In addition to chemical carcinogens, infectious agents like oncogenic viruses are well-recognized causes of neoplasms in other fish species, particularly lymphoid or skin tumors. However, such agents have not yet been identified in zebrafish. The observations from diagnostic cases indicate that older fish are predisposed to higher incidences of most neoplasms. Therefore, removing older fish from colonies should reduce the prevalence of these tumours. Chronic inflammation associated with parasitism is a well recognized as a promoter of neoplasm in mammals and appears to be link to intestinal neoplasm in the zebrafish. In colonies of zebrafish infested with nematodes, intestinal tumours appear to be linked to infections by *Pseudocapillaria tomentosa*. In these cases, eliminating the intestinal nematodes should reduce the occurrence of these neoplasms (Kent and Bishop-Stewart, 2003). N-nitrosodimethylamine (DMN) and N-nitrosomorpholine (NM) cause HCC, cholangiocellular carcinoma (CCC) and oesophageal cancer in zebrafish. In zebrafish, about 8 weeks of exposure, DEN is required to induce the cancer. The fish embryos and fry are both have been found quite responsive to MNNG; however, juvenile zebrafish were remarkably refractory to MNNG-induced neoplasms. Principal target organs in zebrafish treated as embryos with MNNG were the liver and testis, with HCA as the most prevalent hepatic neoplasm. A variety of mesenchymal neoplasms occurred in zebrafish following embryo exposure to MNNG, including chondroma, hemangioma, hemangiosarcoma, leiomyosarcoma and rhabdomyosarcoma. The testis and blood vessels were the primary target organs for MNNG following fry exposure, with seminoma, hemangioma, hemangiosarcoma, and various other epithelial and mesenchymal neoplasms (Madhuri *et al.*, 2012h).

Some of the followings are cancers (or neoplasms) that have been observed in zebrafish during cancer research and in the diagnostic cases (Madhuri *et al.*, 2012h):

Seminoma

This is the most common spontaneous neoplasm which was observed in diagnostic cases. In older broodstock, this neoplasm is often huge in size, causing marked abdominal distention, and constituting about half of the body mass. It is typically soft, white, multilobulated mass, and has always been confined to the testis. The most seminomas of fish are spermatocytic seminomas with extensive differentiation of germ cells into various developmental stages up to sperm. Small seminomas may be comprised of a single lobule of proliferative seminiferous epithelium compressing the surrounding normal testicular tissue. Large seminomas may be multilobular. Various ratios of primordial germ cells to differentiating spermatocytes and spermatids may occur in different seminomas. Most of the seminomas have been seen well differentiated, with histological structures evident that are typically found in normal testes.

Intestinal Carcinoma

This is most commonly recognized only on histology. In extreme cases, a visceral mass may be observed on necropsy which is soft to firm and tan in colour. Intestinal carcinomas occur most commonly in the anterior portion of the intestine, just distal to the oesophagus (zebrafish have no stomach). The capillarid nematode *Pseudocapillaria tomentosa* may be a promoter of this neoplasm. This cancer may vary from small focal, relatively well-differentiated neoplasm forming acini to large highly invasive neoplasm comprised of poorly differentiated small round malignant epithelial cells (small cell carcinoma). Poorly differentiated neoplasms may extensively invade adjacent pancreas and viscera.

Ultimobranchial Neoplasm

The ultimobranchial gland is an endocrine structure that occurs transiently during the development of mammals, but remains a distinct organ in the adult fish and bird. This gland is the counterpart to the thyroid C cells (medullary thyroid tissue) in mammals. Like C cells, the ultimobranchial gland secretes calcitonin to regulate the blood calcium levels. Studies of both diagnostic cases from moribund fish and retired broodstock indicate that the incidence of ultimobrachial neoplasia increases substantially in fish over 1.5 year of age. In most lines of zebrafish, a mass is not seen grossly in fish affected by ultimobranchial tumours, but usually this tumour is detected during histologic studies. Because both seminomas and ultimobranchial cancers are common in male fish over 1.5 year of age, these cancers are generally seen together in older males. Several cases have been found in which the abdomen was grossly distended due to marked enlargement of the spleen in fish with ultimobranchial neoplasia. In such cases, spleen was up to 30X normal size, with cystic degeneration, perhaps due to passive congestion. Ascites also occurred in some cases. The commercial diets seemed prone to develop ultimobranchial cancers at a relatively young age, with some cancers seen by 6 months of age. The normal ultimobranchial gland of the zebrafish is a small ovoid organ located between the heart and oesophagus. Normal gland consists of a cluster of acini lined by tall columnar cells, with basally located nuclei. Neoplastic ultimobranchial glands may be enlarged up to 50X normal size. The neoplastic ultimobranchial cells lose their tall columnar structure and acinar arrangement, forming packets or monotonous sheets of ovoid to polygonal neuroendocrine cells. Malignant neoplasms are composed of less differentiated epithelial cells which invade through the fine fibrous capsule which surrounds the normal gland. Many cases have been seen in which large ultimobranchial carcinomas have invaded into the sinus venosus of heart.

Thyroid Neoplasm

It occurs as soft white, often multilobulated masses on the ventral aspect of lower jaw. It must be differentiated from goiter histologically. Goiter in fish can occur when diets and aquatic environment are low in iodine or when goitrogenic chemicals are present. Goiter can present with very large masses on the ventral jaw and surrounding gills. Histological diagnosis of benign and malignant thyroid neoplasia in fish has been controversial because, unlike thyroid in mammals which is encapsulated, most thyroid follicles occur diffusely in connective tissue surrounding the ventral aorta in lower jaw of fish. In certain fish species, normal thyroid tissue can occur in a variety of locations, including spleen and anterior kidney. The normal thyroid tissue of zebrafish is confined to the vicinity of ventral aorta of lower jaw. Several diagnostic cases of thyroid neoplasm have included the foci of highly anaplastic epithelial cells which are clearly malignant histologically, and in certain cases have invaded many tissues throughout the body. These cases have been confirmed as malignant thyroid adenocarcinomas.

Hepatic Neoplasm

Benign and malignant hepatocellular neoplasms (HCA or HCC) have been the most commonly noticed hepatic neoplasms in diagnostic cases. These neoplasms are soft white to tan masses that may be single or multiple in the liver, and may bulge above the surface of normal hepatic tissue. Liver tumours are frequently seen in fish exposed to chemical carcinogens, in both lab studies or in fish from polluted water. However, liver cancers have also been seen in zebrafish with no known history of exposure to carcinogens. The HCA is the cancer comprised of well-differentiated hepatocytes that maintain their normal orientation into hepatic plates, but which form proliferative masses compressing surrounding hepatic parenchyma. The HCC is comprised of more anaplastic hepatocytes, which have typically lost their normal organization into hepatic plates and often extensively invade into surrounding normal hepatic tissues. These cancers are often associated with foci of hepatocellular alteration. These are clusters of hepatocytes that stain differently from normal hepatocytes like a cytoplasm that is more eosinophilic, more basophilic or more vacuolated due to glycogen accumulation. However, they maintain their normal orientation into hepatic plates and do not compress the surrounding parenchyma. Carcinogenesis studies in mammals and fish indicate that these foci of hepatocellular alteration are often precursors to hepatic neoplasms. Other liver tumours that are not uncommon in fish are those with a biliary origin, *e.g.*, choangiocelluar carcinomas and biliary carcinomas.

Spindle Cell Sarcoma

It may occur in viscera, skeletal muscle, or behind the eye, causing exophthalmia, often as whitish, solid masses. Other tumours (*e.g.*, capillary hemangiomas) may also occur in the eye socket. Fibrosarcomas (malignant neoplasms of fibroblasts), leiomyosarcomas (malignant neoplasms of smooth muscle) and malignant nerve sheath neoplasms are difficult to distinguish from one another grossly, and when poorly differentiated, they may also be difficult to distinguish histologically. Thus, the collective term '*spindle cell sarcoma*' is often used when the precise cell type of the tumour cannot be discerned. Many neoplasms showing cellular architecture consistent with both neural fibrosarcomas and fibrosarcomas have been noticed.

Hemangioma

While most eye tumours that we have seen fall within the collective name '*spindle cell sarcoma*', capillary hemangiomas in the retro-bulbar tissues have also been seen. Diagnosis of neoplasms of zebrafish relies on histological evaluations. The scientists have not yet confirmed the usefulness of immunohistochemical markers for various cell types on formalin-fixed tissues of zebrafish, and thus the histological tumour diagnosis often falls within broad categories.

Transgenic Fish Models for Drug Research

Transgenic Fish Model

As the genetic engineering advances, we are seeing increased commercial application of this technology. Aquatic animals are being engineered to increase aquaculture production, for medical and industrial research, and for ornamental reasons. While some of these alterations may provide some benefits, the potential effects on human health and the environmental risks that *'transgenic fish'* pose to native ecosystems remain unstudied and unknown. *'Transgenic organisms'* can be defined as those which have received sequences of DNA by artificial means, followed by integration of one or more transgenic organisms of the novel sequences into their chromosomal DNA.

In India, research in transgenic fish was initiated in Madurai Kamaraj University (MKU), Centre for Cellular and Molecular Biology, Hyderabad and National Matha College, Kollam with borrowed constructs from foreign scientists. The first Indian transgenic fish was generated in MKU in 1991 using borrowed constructs. India has developed experimental transgenics of rohu fish, zebra fish, cat fish and singhi fish, etc. Genes, promoters and vectors of indigenous origin are now available for only two species namely rohu and singhi for engineering growth. Transgenic rohu produced from indigenous construct at Madurai Kamaraj University has proved to be eight times larger than the control siblings. This transgenic rohu attains 46 to 49 g weight within 36 weeks of its birth. Taking further research to promote transgenic fish programme, the Indian Council of Agricultural Research (ICAR) plans to develop autotransgenesis in commercially important fish species with growth hormone gene (Madhuri *et al.*, 2012b).

Studies on transgenic fish, in contrast to mammals, are still in their infancy. However, it is evident that such fish will not only be of considerable economic benefit to aquaculture but will enable scientists to make quantum leaps in their understanding of the physiological and biochemical mechanisms unique to fish, and of the developmental biology of vertebrates in general. The potential of transgenic fish for research and industrial development is beginning to be widely recognized. Several advantages of fish as animal models have been highlighted, and through an example of a transgenic fish model developed for *in vivo* mutagenesis, the important issues have been elucidated which apply widely to the use of transgenic fish in biological research (Winn, 2001).

Development and Care of Transgenic Fish Models

Although development of new transgenic fish models has increased steadily in recent years, the transgenic fish are not currently being produced as routinely as rodents. Production and perpetuation of lineages of transgenic fish remain a time-consuming process requiring combinations of specialized personnel, equipment and facilities which are not universally available in the laboratories. Procedures and facilities for developing and maintaining transgenic fish are nevertheless undergoing continual improvements to increase the efficiency of the process. Selection of a fish species for development of a transgenic model is dictated by whether the species has certain characteristics that will enable efficient production of the transgenic fish that can be used to address fundamental research questions. Small laboratory aquaria fish species such as medaka, mummichog, and zebrafish share numerous desirable traits for transgenic development and environmental toxicology. Small size, short generation time and cost-effective husbandry contribute to efficient transgenic development and aquarium-based laboratory culture. Well-described embryology, controlled year-round spawning, transparent chorion and short embryogenesis and generation time are advantageous for transgenic production. Medaka has several characteristics that are especially well suited for environmental toxicology, including well-characterized histopathology. They have been used extensively in chemical hazard testing, carcinogenesis bioassays and germ cell mutagenesis studies. As a common inhabitant of coastal environments, the mummichog is one of the most extensively characterized and utilized fish in studies of marine environmental contaminant risk and ecotoxicology (Bunton, 1999).

With increasing numbers of synthetic chemicals introduced into environment each year, concerns remain regarding our understanding of linkages between exposure to toxic agents and potential disease. Consequently, new approaches are needed to assess the health risks associated with exposure to chemical contaminants in aquatic environment. Development of transgenic animal models represents a revolutionary advance in the study of a variety of disease processes. Using introduced genes, rodent genomes are routinely modified to express novel gene products or to over express endogenous gene products. Endogenous genes, such as oncogenes or tumour suppressor genes, can be selectively inactivated or knocked out to produce animals that do not express normal amounts of a specific gene product. Beginning in the late 1980s, researchers started to use genetically modified mice in toxicological studies of chemical carcinogenesis and *in vivo* mutagenesis. Since then, emphasis has continued to be on developing transgenic models to improve and expedite carcinogenicity testing. Development of new transgenic rodent models for measuring other endpoints, such as xenobiotic metabolism and xenobiotic ligand/receptor interactions, is expanding (Mahler, 2000).

Considering the achievements of transgenic rodent models, it is reasoned that the development of transgenic fish could enhance the utilization of fish as indicators of chemical exposure and as non-mammalian animal models in comparative biology. Fish have played significant roles in monitoring and assessing risks of exposure to chemicals in aquatic environments. In some applications, *e.g.*, assessment of toxicity associated with exposure to complex chemical mixtures or in low-dose chronic exposure regimens, fish are recognized as test organisms with distinct and superior benefits in providing insights to disease processes. Fish are also embraced as cost-effective and important animal models in genetics, developmental biology, and toxicology. Development of transgenic fish has been envisioned as a means to enhance the utility of fish models in reducing, refining or replacing selected mammals used in toxicity testing (Winn, 2001).

To expand the utility of fish as environmental sentinel organism (*i.e.*, as indicator of the presence or biological impact of a hazardous substance), transgenic fish have been envisioned carrying reporter genes driven by promoters that are responsive to chemical exposure. In a conceptual approach, the transgenic fish would be placed in the water containing the chemical to be tested. Following uptake, distribution and accumulation of the substance in fish tissues, the genomically integrated response elements would be activated and the reporter genes would be upregulated. The fish would be removed from the water and assayed for reporter gene activity that would be proportional to the concentration of the chemical to which the fish had been exposed. Using this approach, various transgenic fish have been conceived, or are at early stages of development (Carvan *et al.*, 2000).

Several groups have reported difficulties with sustaining gene expression in transgenic lineages, which has slowed the introduction of new transgenic fish. Expression of foreign DNA is commonly observed in the founder fish, but expression is rare or unpredictable in subsequent generations. Suboptimal expression may result from various factors, including the use of heterologous transcription units derived from non-fish species, from methylation of CpG dinucleotides, from incorporation of multiple transgenes into a single site, or from integration of transgenes in loci susceptible to gene silencing. Examples include fish carrying metal-responsive or heat-shock promoters spliced into green fluorescent or LUC reporter genes to provide an *in vivo* indicator of exposure to heavy metals or other pollutants. As more groups gain experience using a variety of methods, such as insulating border elements to stabilize the expression of transgenes, it is anticipated that new fish models with improved transgene expression will be introduced. A promising example is demonstrated by a transgenic fish assay recently introduced to determine the effects of oestrogenic chemicals at critical life stages on sensitive target organs in fish (Winn, 2001).

Legler *et al.* (2000) introduced an estrogen binding sequence linked to a TATA box and a luciferase reporter gene in zebrafish (*D. rerio*). Binding of the chemical substance to endogenous estrogen receptors and subsequent transactivation of the estrogen receptors (ERs) induced the luciferase gene as measured in tissue lysate. The study revealed that the period of gonad differentiation was highly sensitive in juvenile fish exposed to oestradiol (an oestrogen). In adult males, the testis was the target tissue most sensitive and responsive to oestrogens.

In conclusion, the transgenic fish are produced through the introduction of foreign DNA sequences into zygotes or embryos resulting in the genomic integration of the DNA in a stable and heritable manner. Manual injection of DNA using drawn capillary needles via the cytoplasm of newly fertilized eggs remains the most commonly used method of introducing DNA into the fish genome. A variety of methods have also been used, including microinjection of the germinal vesicle, electroporation of embryos or sperm, retroviral infection and particle gun bombardment (Winn *et al.*, 2001).

Transgenic Fish Mutation Models

Several transgenic fish models have been produced that carry prokaryotic vectors harboring specific genes, which serve as targets for quantifying spontaneous and induced *in vivo* mutations. Common approach to analyze the mutations in transgenic rodent and fish models entails treating the animals with a chemical and allowing sufficient time for manifestation of the mutations. Genomic DNA is then isolated from various tissues, and vectors are separated and recovered upon transfer into specialized indicator bacteria where the mutant and non-mutant target genes are readily distinguished and quantified. Transgenic mutation assays provide numerous benefits for analyzing mutations *in vivo* not available using other approaches. A significant challenge to detecting mutations in whole animals is the practical problem of recovering and identifying mutant genes. Transgenic mutation assays afford efficient recovery and screening of large numbers of copies of a transgene target, providing statistically meaningful results as well as reducing the numbers of animals required in a chemical treatment. Typically, hundreds of thousands of mutation target loci are recovered from as little as 1 to 5 µg of DNA and as few as 6 to 10 animals/treatments are required to detect significant induction of mutations above a background mutation frequency (Winn, 2001).

Mutations in transgenic assays are detected in genetically neutral targets, thereby avoiding potential selective pressures on the mutant frequency *in vivo*, and allowing the accumulation and persistence of mutations. As a consequence of the accumulation of mutations in these loci over time, repeated or chronic chemical treatments will increase the sensitivity of the mutation assay. The amenability of fish to a wide range of chemical treatment regimens indicates that fish will be well suited to mutation studies using treatments that closely approximate environmental exposure conditions. Transgenic medaka was produced which carried multiple copies of bacteriophage vector, that harbours bacterial genes as mutational targets. By adapting a mutagenesis system originally developed for rodents to fish, the process of transgenic fish model development was made more efficient and increased the potential utility of fish model for comparative mutagenesis studies. Different mutants were also recovered with high efficiency from a variety of fish tissues, including whole fish, liver, testis, eye, whole blood and skin. This ability to compare mutations among tissues provides an important tool for understanding differential responses to mutagens. The mutagen exposure studies illustrated the importance of mutation manifestation time, or interval between mutagen treatment and analyses. The time required for mutations to manifest is affected by several variables, including tissue/cell type, mutagen and mutagen treatment regimen, which must be considered in designing and interpreting mutation studies. Furthermore, fish were exposed to DMN, a potent liver carcinogen which requires cell proliferation for DMN-produced methyl DNA adducts to become fixed as mutations (Winn (2001). Cell proliferation is a requisite for DNA repair and mutation (Bielas and Heddle, 2000).

Winn (2001) developed a transgenic fish mutation assay based on a bacteriophage vector using medaka (*O. latipes*) and mummichog (*F. heteroclitus*) carrying bacteriophage vector. This worker was able to show that the spontaneous mutant frequencies in fish are comparable with those of transgenic mice carrying the identical vector. This mutation assay has not had wide use; however, results demonstrated the feasibility of generating transgenic mummichog, which should encourage expanded development of transgenic marine species as models for environmental studies. Amanuma *et al*. (2000) adapted to zebrafish a mouse mutation assay. They showed that the spontaneous mutant frequency in zebrafish is similar to transgenic mice containing the same target, and they observed a dose-dependent induction of mutants after treatment of fish embryos with ethylnitrosourea. Another plasmid-based mutation system adapted from a rodent mutation assay is in early stages of development. The mutation assay is based on transgenic medaka and mummichog, which carry the plasmid vector harbouring the gene as a mutation target. After exposure of the animal to a mutagen, genomic DNA is isolated and digested with a restriction enzyme to release monomeric plasmid sequences. Analyses revealed that the plasmid vector could be recovered from fish tissues with exceptional efficiency; the spontaneous mutant frequency in the fish was similar to that of the transgenic rodent carrying the identical transgene (Winn, 2001).

Researchers have developed mutation assays that are not reliant on the expression of a transgene in the animal to assess potential DNA damage after environmental chemical exposure. Using procedures similar to that developed for rodents, scientists also produced several transgenic fish models which carry prokaryotic vectors harboring specific genes that serve as targets for quantifying spontaneous and induced *in vivo* mutations. Common approach to analyzing mutations in transgenic rodent and fish models entails treating the animals with a chemical and allowing sufficient time for manifestation of mutations. Genomic DNA is then isolated from various tissues, and the vectors are separated and recovered upon transfer into specialized indicator bacteria, where mutant and non-mutant target genes are readily distinguished and quantified. Transgenic mutation assays provide numerous benefits for analyzing mutations *in vivo* not available using other approaches. A significant challenge to detecting mutations in whole animals is the practical problem of recovering and identifying mutant genes. Need to detect mutations at very low frequencies demands highly efficient means of recovering and distinguishing mutant genes among a very large number of non-mutant genes. Transgenic mutation assays afford efficient recovery and screening of large numbers of copies of a transgene target, providing statistically meaningful results as well as reducing the numbers of animals required in a chemical treatment. Transgenic mutation assays facilitate comparisons of mutational responses among identical loci recovered from different cells, tissues, organs and species. In contrast to assays based on endogenous genes in which mutations are detected only in specific developmental stages or specific tissues, mutations in transgene targets can be examined in virtually any tissue from which DNA may be isolated.

Mutation analyses can also be combined with measures of other endpoints like sequencing of specific mutations to aid in disclosing possible mechanisms of mutagen action. Utilization of transgenic mutation models in a broad range of comparative analyses is improving our understanding of the influences of cell proliferation, metabolism, toxicity and DNA repair in mutagenesis. Before the advent of *in vivo* assays with the capability of detecting mutations directly in transgenic targets, researchers relied on genetic toxicity assays that focused on endpoints other than the mutation. Among these endpoints, researchers have used induction of DNA adducts, DNA repair, DNA strand breakage and chromosomal damage as indices of genotoxicity *in vivo* in fish. These analyses have provided valuable information regarding exposure of organisms to genotoxic agents and indirect information on alterations of DNA or chromosome structure induced by a chemical. However, these methods have limited sensitivity in detecting mutations at the DNA level, the ultimate endpoint of DNA damage and/or attempted repair. Mutations in transgenic assays are detected in normally neutral targets, thereby avoiding potential selective pressures on the mutant frequency *in vivo* and allowing accumulation and persistence of mutations (Madhuri *et al.*, 2012b; Winn, 2001).

As a consequence of the accumulation of mutations in the loci overtime, repeated or chronic chemical treatments will increase the sensitivity of mutation assay. Amenability of fish to a wide range of chemical treatment regimens indicates that fish will be well suited to mutation studies using treatments that closely approximate environmental exposure conditions. Although development of new transgenic fish models has increased steadily in recent years, transgenic fish are not currently being produced as routinely as rodents. Production and perpetuation of lineages of transgenic fish remain a time-consuming process requiring combinations of specialized personnel, equipment and facilities, which are not universally available in research laboratories. Procedures and facilities for developing and maintaining the transgenic fish are nevertheless undergoing continual improvement to increase the efficiency of this process. Selection of a fish species for development of a transgenic model is dictated by whether the species has certain characteristics, which will enable efficient production of transgenic fish that can be used to address fundamental research questions. Small laboratory aquaria fish species like medaka, mummichog and zebrafish share numerous desirable traits for transgenic development and environmental toxicology. Small size, short generation time and cost-effective husbandry contribute to efficient transgenic development and aquarium-based laboratory culture. Well-described embryology, controlled year-round spawning, transparent chorion, and short embryogenesis and generation time are advantageous for transgenic production. Medaka fish has several characteristics, which are especially well suited for environmental toxicology, including well-characterized histopathology. They have been used extensively in chemical hazard testing, carcinogenesis bioassays and germ cell mutagenesis studies. As a common inhabitant of coastal environments, the mummichog is one of the most extensively characterized and utilized organisms in studies of marine environmental contaminant risk. Transgenic fish are produced through the introduction of foreign DNA sequences into zygotes or embryos resulting in the genomic integration of the DNA in a stable and heritable manner. Manual injection of DNA using drawn capillary needles via the cytoplasm of newly fertilized eggs remains the most commonly used method of introducing DNA into the fish genome. A variety of methods have also been used, including micro-injection of the germinal vesicle, electroporation of embryos or sperm, retroviral infection and particle gun bombardment (Madhuri *et al.*, 2012b; Winn, 2001).

Establishing and Maintaining Transgenic Fish Lineages

Transmission of a transgene through the germ line is a requisite for most transgenic animal studies. Whereas the process of gene transfer can be technically challenging, the identification of transgenic founders, performance of appropriate crosses and perpetuation of desired lineages are not technically difficult. However, these tasks can burden resources and personnel. Mosaic integration of a transgene in the founder animals (*i.e.*, integration in only a portion of the animal's cells) is the rule in transgenic fish. Although the presence of transgene may be demonstrated in excised fin tissue, the transgene is often not transmitted, or is transmitted only at a low frequency to the next generation. In addition, the need to identify transgenic fish that meet the specific experimental requirements, places extra demands on the process of establishing lineages. The problems may arise relating to integration or function of a transgene, or the animal may have poor reproduction or an undesirable phenotype, thereby reducing the utility of lineage. It is advisable to establish the multiple lineages to allow for these differences (Madhuri *et al.*, 2012b; Winn, 2001).

Fish researchers, being most familiar with the requirements for care, handling and maintenance of laboratory fish, are typically the primary animal care providers. Similarly, other researchers may recognize the value of fish as models and may want to pursue the production of transgenic fish but may not have necessary experience with fish husbandry or transgenic fish technology. To assist in promoting optimal use of fish models and to bring standards, practices and facilities for care and utilization of fish up to that of mammalian models, various improvements are needed in aquatic animal resource programmes. In particular, improvements in training of personnel in fish care and transgenic techniques, increased research on diet, disease diagnosis and prevention, and increased institutional support of aquatic research programmes will aid in expanding and developing new models and ensure continuity in long-term care. Before embarking on the development of transgenic fish, researchers are advised to address issues regarding assessment of the potential risks associated with environmental release of transgenic fish.

It may be argued that most small aquaria fish species are especially dependent on the laboratory environment for survival. For most transgenic models developed for applications in environmental toxicology, fish would conceivably carry new genes that are either neutral or deleterious to animal's health. Nonetheless, it is advisable that procedures and facilities used to maintain transgenic fish be reviewed in accordance with guidelines proposed for development, care and use of genetically modified fish and shellfish to ensure that precautions are taken to safeguard against releases of transgenic fish into the environment (Madhuri *et al.*, 2012b; Winn, 2001).

Use of Transgenic Fish in Pharmacological and Toxicological Research

Scientists have generated numerous transgenic fish using a variety of species and transgenes since the first transgenic fish were introduced in 1985. Despite wide recognition of fish as sensitive subjects for chemical toxicity testing and as valuable comparative animal models, the field of environmental toxicology has only recently begun to benefit from transgenic technology. The emphasis for transgenic fish research has been on refining transgenic methodology or developing novel strains of fish with commercially beneficial traits. However, recent developments indicate that transgenic fish are promising as new research animal models (Carvan *et al.*, 2000).

Transgenic animal models hold significant promise for enhancing studies on the genetic and molecular basis of disease and for improving the scientific basis for environmental health risk assessment. Studies from transgenic fish can yield reliable and reproducible data, using precisely defined pharmacological/toxicological endpoints while reducing costs and numbers of animals. By taking full advantage of amenability of fish to a variety of manipulations, ranging from microinjection of embryos to static-renewal and flow-through chronic exposures, transgenic fish can improve the assessment of realistic risks related to exposure to waterborne and sediment-associated contaminants. As the utilization of transgenic fish models intensifies, it is anticipated that transgenic fish will contribute significantly to emerging issues related to environmentally induced, reproductive, developmental and heritable diseases for which such studies in mammalian models are exceedingly difficult (Madhuri *et al.*, 2012b; Winn, 2001).

Attraction of transgenic fish models is moderated by the uncertainty of how these new models will be applied most effectively in environmental toxicology/pharmacology. Limited availability of lineages of fish beyond laboratories from which they were generated and limited databases on responses under a variety of test conditions will potentially slow wide adoption by the researchers. For each new model and application, it will be necessary to resolve complex issues related to establishing optimal study designs like appropriate exposure regimens, chemical concentrations, routes and duration of exposures, and number of animals required for a test. Besides, requirement for reproducibility of responses in test animals will demand rigid husbandry practices, many of which are not standardized among different laboratories, to reduce variability in culture conditions. Guidance in addressing some of these issues may be obtained from examples provided by application of transgenic rodent models in toxicology (Mahler, 2000; Winn, 2001).

Chapter 3
Pharmacological Regimens of Various Fish Diseases

Chemotherapy Regimens and Chemotherapy

Meaning and Definition

'*Chemotherapy regimen*' is one of the pharmacological regimens employed for the treatment of different specific diseases of humans as well as animals, including fish. Hence, '*chemotherapy regimens*' are regimens for '*chemotherapy*' that combine several '*chemotherapy drugs*'. These drugs are natural, semisynthetic, or synthetic substances with selective inhibitory effects against biological pathogenic agents (microorganisms) of humans and animals (also fish), and against atypical (cancerous) cells.

On the other hand, '*chemotherapy*' is the use of medications (chemicals) to treat disease. It can include the '*antibiotics*' or other medications to treat any illness or infection. Chemotherapy in its most general sense is the treatment of disease by chemicals, especially by killing microorganisms or cancerous cells. In popular usage, it refers to '*antineoplastic drugs*' used to treat cancer; or combination of these drugs is used into a cytotoxic standardized treatment regimen. In its non-cancerous use, chemotherapy refers to '*antibiotics or antibacterial chemotherapy*'. In relation to treatment of cancer, chemotherapy (also abbreviated as '*chemo*') is the treatment of cancer with one or more chemotherapeutic agents (or cytotoxic antineoplastic drugs) as part of a standardized regimen. In other words, '*chemotherapy*' typically refers to the destruction of cancer cells. It may be given with a curative intent, or it may aim to prolong the life or to palliate the symptoms. An older and broader usage of the term '*chemotherapy*' encompassed any chemical treatment of disease (*e.g.*, treatment of infections by antimicrobial agents). However, this usage has become archaic. The word *'chemotherapy'* without a modifier usually refers to *'cancer treatment'*, but its historical meaning is broader (Pandey, 2013f).

A fundamental philosophy of oncology, including combination chemotherapy, is that different drugs work through different mechanisms, and that the results of using multiple drugs will be synergistic to some extent. Because they have different dose-limiting adverse effects, they can be given together at full doses in chemotherapy regimens. The first successful combination chemotherapy was *'MOPP'* introduced in 1963 for lymphomas. MOPP includes *'mustargen'* (also known as mechlorethamine, mustine, nitrogen mustard or MSD), *'oncovin'* (also known as vincristine), *'procarbazine'* (also known as matulane or natulan) and *'prednisone'* (also known as deltasone or orasone). MOPP is a combination chemotherapy regimen used to treat Hodgkin's disease. The acronym is derived from the component drugs of the regimen.

The term *'induction regimen'* refers to a chemotherapy regimen used for the initial treatment of a disease. A *'maintenance regimen'* refers to the ongoing use of chemotherapy to reduce the chances of a cancer recurring or to prevent an existing cancer from continuing to grow. Chemotherapy regimens are often identified by acronyms, identifying the agents used in the drug combination. However, the letters used are not consistent across regimens, and in some cases- for example, *'BEACOPP'*- the same letter combination is used to represent two different treatments. There is no widely accepted naming convention or standard for the nomenclature of chemotherapy regimens, *e.g.*, either genetic or brand names may be used for acronyms.

Categories of Chemotherapeutic Drugs

Most of chemotherapeutic drugs are divided into following classes (Pandey, 2013f):

1. Antibacterial (or antimicrobial) drugs
2. Antitubercular drugs
3. Antileprotic drugs
4. Antiviral drugs
5. Antifungal drugs
6. Antiprotozoal drugs
7. Antihelmintic drugs (or anthelmintics)
8. Antineoplstic (or anticancer/cytotoxic/cytostatic) drugs

The first modern chemotherapeutic agent was *'arsphenamine'* (an arsenic compound) discovered by Paul Ehrlich in the year 1909. This drug was used to treat syphilis. This was later followed by the *'sulphonamides'* (sulpha drugs, discovered by Domagk) and *'penicillin'* (discovered by Alexander Fleming in 1928-29). The use of minerals and plant-based medicines is believed to date back to prehistoric medicine. Therefore, in the 1940s, several patients with advanced lymphomas (cancers of certain white blood cells, WBCs) were given the drug by vein, rather than by breathing the irritating gas. Their improvement, although temporary, was remarkable. That experience led researchers to look for other substances that might have similar effects against the cancer. As a result, many other drugs have been developed to treat cancer, and the drug development since then has exploded into a multibillion-dollar industry, although the principles and limitations of chemotherapy discovered by the early researchers still apply. After a military operation in the World War II, some sailors were accidentally exposed to mustard gas. They were later found to have very low WBC counts. The WBCs usually grew very quickly and the cancer cells also divided and grew very quickly. The doctors wondered whether the effect of mustard gas-slowing down the rapid growth of WBCs- may have same effect on the cancer cell growth. Now, there are more than 100 different types of chemotherapeutic drugs, which can treat the most cancers (Pandey, 2013f).

Action of Chemotherapy

Chemotherapy mostly acts by killing the cells which divide rapidly, one of the main properties of cancer cells. It means that it also harms cells which divide rapidly under the normal circumstances: cells in the bone marrow, digestive tract and hair follicles. The newer anticancer drugs act directly against abnormal proteins in the cancer cells; this is termed as *'targeted therapy'*. Chemotherapy drugs attack cells in the process of growth and division. Individual drugs may work through many different mechanisms, such as damaging a cell's genetic material (DNA), or preventing the cell from dividing. However, the chemotherapeutic drugs cannot distinguish between malignant cancer cells and normal cells. All rapidly dividing cells are potentially sensitive to chemotherapy. Cytotoxic drugs may lead to complete remissions for some disseminated cancers (*e.g.*, lymphoma), may be effective in decreasing tumour size and may prolong the life in many other types of metastatic cancers like osteosarcoma. Choice of specific therapies depends on tumour type (what it is), histologic grade of tumour (how aggressive it is), stage of disease (where it is) and patient's (but mainly the pet owner's) tolerance for the side effects of various treatments. Most of the chemotherapy protocols designed for veterinary patients have <5 per cent incidence of severe, life-threatening complication. Most veterinary chemotherapy patients enjoy a good quality of life while on therapy. The *'cytotoxic antibiotics'* are a varied group of drugs which have various mechanisms of action. This class of drugs includes anthracyclines and other drugs, including actinomycin, bleomycin, plicamycin and mitomycin. Doxorubicin and daunorubicin were the first two anthracyclines (obtained from the bacterium *Streptomyces peucetius*). The derivatives of these compounds are epirubicin and idarubicin. Other clinically used drugs in the anthracyline group are pirarubicin, aclarubicin and mitoxantrone. The mechanisms of anthracyclines include DNA intercalation (molecules insert between the two strands of DNA), generation of highly reactive *'oxygen free radicals'* (which damage the intercellular molecules) and topoisomerase inhibition. Actinomycin is a complex molecule which intercalates DNA and prevents RNA synthesis. Bleomycin (a glycopeptide isolated from *Streptomyces verticillus*) also intercalates DNA, but produces free radicals that damage DNA. This occurs when bleomycin binds to a metal ion, becomes chemically reduced and reacts with oxygen. Mitomycin is a cytotoxic antibiotic with the ability to alkylate DNA (Pandey, 2013f).

Uses and Dosage of Chemotherapy

Chemotherapy was historically used for non-oncological (non-cancerous) references, such as the use of antibiotics (antibacterial chemotherapy). But, it is often used in conjunction with other cancer treatments, *e.g.*, radiation therapy or surgery. Certain chemotherapeutic agents also have a role in the treatment of other conditions, including ankylosing spondylitis, multiple sclerosis, Crohn's disease, dermatomyositis, psoriasis, psoriatic arthritis, systemic lupus erythematosus, rheumatoid arthritis, polymyositis, scleroderma and suppression of transplant rejections. Some newer anticancer drugs (*e.g.*, various monoclonal antibodies) are not indiscriminately cytotoxic, but rather target proteins that are abnormally expressed in cancer cells and that are essential for their growth. Such treatments are often referred to as *'targeted therapy'* (as distinct from *'classic chemotherapy'*), and are often used alongside traditional chemotherapeutic agents in antineoplastic treatment regimens. Chemotherapy is likely to be recommended for cancer that has already spread to other areas of the body (metastatic disease), for tumours that occur at more than one site (multicentric disease), or for tumours that cannot be removed surgically (non-resectable disease). In some cases, chemotherapy can be used to try to shrink the large tumours prior to surgery or to help eradicate certain types of microscopic cancer cells that cannot or have not been completely removed surgically. For cancers that are at high-risk for metastasis early in the course of disease, chemotherapy can be used after surgery or radiation therapy to help slow down the growth of cancer cells in other parts of the body. In animals also, chemotherapy is used to treat systemic cancers, that have already spread or metastasized, or the cancers that are likely to spread or metastasize. Sometimes, chemotherapy can be used to treat the cancers that cannot be treated with other methods like surgery or radiation (Pandey, 2013f).

Dosage of chemotherapy can be difficult. If the dose is too low, it will be ineffective against the tumour; whereas, at excessive doses, toxicity (side effect) will be intolerable to the patient. This has led to the formation of detailed *'dosing schemes'*, which give guidance on the correct dose and adjustment in case of toxicity. In immunotherapy (treatment of autoimmune disorders), chemotherapeutic agents are used in smaller dosage than in the treatment of malignant diseases. Most chemotherapy is delivered intravenously, although a number of agents can be administered orally (*e.g.*, melphalan, busulfan, capecitabine). There are many intravenous (iv) methods of drug delivery, known as *'vascular access devices'*. These include the winged infusion device, peripheral cannula, midline catheter, peripherally inserted central catheter, central venous catheter and implantable port. The devices have different applications regarding duration of chemotherapy treatment, method of delivery and types of the chemotherapeutic drugs. For continuous, frequent or prolonged iv chemotherapy, different systems may be surgically inserted into the vasculature to maintain access. In animals, chemotherapy can be given intravenously (the most common method), intramuscularly, subcutaneously or orally for cancer. Some drugs are oral medications (pills) which can be given at home. Others are brief injections that require an outpatient appointment. In some instances, slow infusions or repeated treatments throughout the day may require. Treatments are typically repeated weekly for every third week. Blood tests may be needed to monitor the effects of chemotherapy during the weeks between the treatments. Duration of chemotherapy depends on the type of cancer and extent of disease. Some animals need to receive the chemotherapy for the rest of their lives. In others, the treatments may be spread out or discontinued after a period of weeks to months provided that cancer is in remission, *i.e.*, there is no detectable evidence of cancer in the body. Chemotherapy can be resumed when cancer relapses. Usually, every patient should receive at least two cycles of chemotherapy, and then be evaluated for response before continuing the treatment, changing the drugs or discontinuing the chemotherapy (Pandey, 2013f).

Remedy of Common Fish Diseases

Common Fish Diseases

With increasing numbers of synthetic chemicals introduced into the environment each year, concerns remain regarding our understanding of the linkages between exposure to toxic agents and potential disease. Chemical contaminants of aquatic environments is of significant concern because although it is understood that aquatic systems serve as major conduits for distribution and deposition of many toxic agents, relatively few methods are available which provide sufficient sensitivity, accuracy and practicality necessary for assessment of chemical toxicity. As a consequence, new approaches are needed to improve the assessment of health risks associated with exposure to chemical contaminants in the aquatic environments (Madhuri *et al.*, 2012b).

Similar to other animals, fish can also suffer from various types of diseases. All fish carry pathogens and parasites. Usually this is at some cost to the fish. If the cost is sufficiently high, then the impacts can be characterized as a disease. However, disease in fish is not understood well. What is known about fish disease often relates to aquaria fish, and more recently, to farmed fish. Disease is a prime agent affecting fish mortality, especially when fish are young. Fish can limit the impacts of pathogens and parasites with behavioural or biochemical means, and such fish have reproductive advantages. Interacting factors result in low grade infection becoming fatal diseases. In particular, things that cause stress, such as natural droughts or pollution or predators, can precipitate outbreak of disease. Disease can also be particularly problematic when pathogens and parasites carried by introduced species affect native species. An introduced species may find invading easier if potential predators and competitors have been decimated by disease. Pathogens which can cause fish diseases comprise: virus, bacteria, fungi, mould and water infections, etc. (Madhuri *et al.*, 2012i).

Fish are exposed from various pollutants, including drugs and chemicals. Fish can also be infected by different pathogens, microorganisms or parasites. Most common fish diseases, particularly in freshwater aquariums, include columnaris, gill disease, ich, dropsy, tail and fin-rot, fungal infections, white spot disease, pop-eye, cloudy eye, swim bladder disease, lice and nematode worms infestation, water quality induced diseases, constipation, anorexia, chilodonella, ergasilus, tuberculosis, glugea, henneguya, hexamita, hole-in-the-head disease (head and lateral line erosion disease, lateral line erosion or lateral line disease), injuries, leeches in aquariums, lymphocystis, marine velvet, and neon-tetra disease (Pandey, 2013c; Pandey *et al.*, 2012b). Bacterial infections are major cause of mortality in aquaculture. Among the common fish pathogenic bacteria, *S. agalactiae, L. garvieae, E. faecalis* (all Gram-positive), and *A. hydrophila* and *Y. ruckeri* (both Gram-negative) cause fish diseases (Pandey *et al.*, 2012b).

Defence by Fish Against Disease

Diseases in fish caused by bacteria are most widespread. Antibiotics are frequently used to control such fish diseases, but there is an increasing risk of developing antibiotic resistant strains of bacteria. The non-specific immune functions like bacteriolytic activity and leukocyte function of fish have been improved by some herbs (Pandey *et al.*, 2012b).

Treatment of bacterial diseases with different herbs has been safely used in organic agriculture, veterinary and human medicine. The herbs mitigate many of the side effects which are associated with synthetic antimicrobials. In addition, the plant-derived drugs provide a cheaper source for treatment and greater accuracy than chemotherapeutic agents. Plants have been used as traditional medicine since time immemorial to control bacterial, viral, fungal and other diseases. Now, research has been initiated to evaluate the feasibility of herbal drugs in fish diseases (Madhuri *et al.*, 2012a).

Fish have a variety of defenses to prevent the diseases and parasites. *'Non-specific defence'* includes the skin and scales, as well as the mucus layer secreted by the epidermis which trap and inhibit the growth of microorganisms. If pathogens breach these defenses, fish can develop an inflammatory response, which increases blood flow to the infected region and delivers WBCs that attempt to destroy pathogens. The *'specific defence'* responses to particular pathogen recognized by the fish's body, *i.e.*, to an immune response. Nowadays, vaccines have become widely used in aquaculture and also with ornamental fish, *e.g.*, furunculosis vaccine in farmed salmon and koi herpes virus vaccine in koi. Some species of fish use the *'cleaner fish'* to remove external parasites. The best known of these is the *'blue streak'* of genus *Labroides* found on coral reefs in the Indian and Pacific oceans. These small fishes maintain so-called the *'cleaning stations'*, where other fishes congregate and perform specific movements to attract the attention of cleaners. Cleaning behaviours have been observed in a number of fish groups, including an interesting case between two cichlids of the same genus, *Etroplus maculatus* (cleaner) and the much larger fish, *Etroplus suratensis* (Madhuri *et al.*, 2012i).

Common Viral Diseases of Fish and their Treatment

The most common viral diseases of fish are as follows (Madhuri and Pandey, 2014; Pandey *et al.*, 2015):

1. Spring Viraemia or Spring Viraemia of Carp (SVC)

Rhabdovirus carpio is the causal agent (pathogen) of this disease, and the disease is mainly restricted to European countries. Common carp is the host species for this disease. The infected fish becomes black and develops legions on the skin and gills.

Bleeding from scales, accumulation of fluid in the body and inflammation of alimentary canal are some other symptoms.

2. Fish Pox

This is also reported in European countries. The epidermis of the infected carp becomes proliferous, that is why the lesions or blisters are formed on the skin.

3. Viral Haemorrhagic Septicaemia (VHS)

This disease is caused by *'viral haemorrhagic septicemia virus'* (VHSV). In this, *'telltale haemorrhages'* appear as red spots on the fish bodies (Figure 37). Severe glomerular changes are seen, resembling a membranous glomerulonephritis with focal necrosis and degeneration in the kidney. The liver sinusoids become engorged with blood along with widespread necrosis.

Figure 37: VHS with Telltale Haemorrhages (Red spots) on Bodies of Pacific Sardines.

4. Infectious Pancreatic Necrosis (IPN)

IPN is caused by *'infectious pancreatic virus'* (IPV). In this disease, the intestine becomes swollen, filled with mucus and catarrhal exudates, especially in salmon parr fish (Figure 38). Focal necrosis of acinar pancreatic tissue is replaced by a loose fibrous network and fatty degeneration. Macrophages and WBCs may infiltrate pancreatic and hepatic tissues. There may be necrosis and sloughing of caecal endothelium.

Figure 38: IPN with Swollen Intestine of Salmon.

5. Lymphocystis

This is a common viral disease of fresh and salt water fishes, caused by *'lymphocystis viruses'*. If a fish has growths resembling raspberries, it may be infected with lymphocystis. The aquarists usually come across this virus when their fish are stressed, *e.g.*, when put into a new environment and poor water quality, the virus is able to grow. The fish starts growing small white pin-prick or tumour like growths on their skin or fins. So, this disease is often mistaken for ich (ick, caused by *Ichthyophthirius multifiliis* protozoan ciliate) in the early stages. It soon clumps together to form a cauliflower-like growth on the skin, fins, and occasional gills. Eventually, such growths inhibit the fish's ability to swim, breathe or eat, and the secondary bacterial infections usually kill the fish. Disease can be inherited by the parent fish or transmitted to other fish through abrasions on the skin.

Although there are no specific treatments for viral diseases of fish, however, certain vaccines and antibiotics can be given in some cases of fish viral diseases.

Common Bacterial Diseases of Fish and their Treatment

Fish are susceptible to several bacterial infections, mainly when reared in high density conditions. Disease outbreaks elevated the mortality rate and decrease the productive efficiency, causing high economic loss of the fish farmers (Madhuri *et al.*, 2012a). *S. agalactiae*, *L. garvieae* and *E. faecalis* are closely related groups of bacteria that can cause diseases like streptococcosis, lactococcosis, haemorrhagic septicemia

(HS) and ulcers in fins. *Flavobacterium columnare* is pathogenic only to freshwater fish and shows low environmental fitness when compared with other aquatic bacteria. Even though, this agent is highly virulent to young fish (fry and fingerling), causing skin lesions and high mortality, usually associated with poor environmental conditions. Enteric red mouth disease mostly restricted to salmonids is caused by *Y. ruckeri*, and reddening of mouth and throat is the most common symptom. *A. hydrophila* is responsible for skin infections, septicemia and gastroenteritis in fish and human. *A. hydrophila*, the most common bacterial pathogen in freshwater fish, is and aetiological agent of many distinct pathological conditions including tail/fin rot, motile Aeromonas septicemia (MAS) or HS and epizootic ulcerative syndrome (EUS) as a primary pathogen. EUS is a globally distributed disease and is an epidemic, affecting a variety of wild and cultured fish species, especially in Southeast Asia, including Pakistan and India (Abdul Kader Mydeen and Haniffa, 2011; Pandey *et al.*, 2012b).

The most common bacterial diseases of fish are as follows (Madhuri and Pandey, 2014; Pandey *et al.*, 2015):

1. Fin Rot or Fin and Tail Rot Disease (Fin Rot and Tail Rot Disease)

It is caused by *'fin rot bacteria'* (FRB), which attacks fins and in extreme cases, the body. FRB is usually the result of environmental stress like poor water quality, bullying, etc. If this left untreated, secondary infections can occur. *'Fin or tail rot'* can also be a secondary infection caused by stress from other diseases. This disease affects both adults and young fishes. The infection during its early stage appears as a white line on the margin of the fin, spreading and imparting frayed appearance to the appendage which eventually putrefies and disintegrates. The disease is contagious and may cause considerable damage. The bacteria usually do not affect healthy fish. When fish are stressed, or when the fins have already been damaged due to excessive fin nipping, then the disease takes root.

2. Ulcer Disease or Columnaris Disease

This disease, caused by *Flavobacterium columnare* (earlier known as *Flexibacter columnaris*) bacteria, is mostly a direct result of poor water quality. Sometimes, the shock of being introduced to the new aquarium leads to this disease in newly imported fish. It shows raised white plaques, often with reddish peripheral zone leading to haemorrhagic ulcers (Figure 39). Grayish-white marks or patches on the body of fish or around its mouth are the first obvious signs. These patches appear like threads, especially in the vicinity of the mouth. Thus, it is generally confused with *'mouth fungus disease'*. The fish fins may show signs of deterioration, the gills get affected and slowly sores start appearing on the body. In later stage, the fish can hold its fins very close to the body and not spreading them. The disease will, however, invade the internal organs too during the later stages.

3. Dropsy

This condition (Figure 40) is not a specific disease, but rather a symptom of a deteriorated health condition. The dropsy bacteria affect to fish weakened by bad water conditions, stress, etc. Accumulation of fluid inside the body cavity, scale

Figure 39: Columnaris Disease.

Figure 40: Dropsy in Goldfish.

protrusion, exophthalmic condition, inflammation of intestine and haemorrhagic ulcers occur on the skin and fins. As a fish not being able to regulate the amount of fluid in a part of its body, it will have visible swelling and projected scales. The affected area is typically the abdomen; specifically, it is most often the visceral cavity that houses a number of organs, *e.g.*, stomach, intestine, gall bladder and kidney. The failure to regulate fluids is a symptom; therefore, there is usually some other disease involved that starts the process (caused by poor water quality, stress, internal bacterial infections, parasites, viruses and tumours). Although dropsy is fairly easy to diagnose, the cause is much harder to determine; however, the primary cause is usually attributed to a bacterial infection. The causative agent can be introduced to the aquarium through food, poor water quality or through the introduction of other fish to an established aquarium. Although dropsy is not highly contagious, the affected fish should be removed and placed in a quarantine aquarium. This disease can be spread from the affected fish, which can possibly produce stress among the other fish and make them more vulnerable to dropsy or other diseases. Main symptoms are bloating, protruding scales, exhaustion, isolation and loss of appetite.

4. Eye Disease

This is an epidemic eye disease which affects medium-sized and large sized catla fish. It is caused by a variant of the bacterium, *Aeromonas liquefaciens*. The infected sites are eyes, optic nerves and brain of the fish.

5. Pop-eye

This is normally caused by bacterial septicemia, tuberculosis, parasites or as a result of oxygen supersaturation of water. Oxygen supersaturation occurs whenever pressure of a gas in water is higher than the pressure of the same gas in surrounding atmosphere; the difference in gas pressures causes the gas to get pulled too quickly out of the fish's bloodstream, leaving behind the gas bubbles. Poor water quality, injury, poor nutrition, tumours and vitamin deficiency can also cause pop-eye. In this, eyes protrude from the head/socket, and may appear inflamed and cloudy.

6. Tuberculosis

It is also called *'fish tuberculosis, piscine tuberculosis, acid-fast disease or granuloma disease'*. It is caused by Gram-positive mycobacteria, *Mycobacterium marinum* or *M. fortuitum*. It is a chronic, progressive disease that may take years to fully develop. If a fish has a bent or curved spine, it is most likely suffering from tuberculosis. Symptoms include lethargy, emaciation, fin and scale loss, exophthalmia (bulging eyes), skin inflammation and ulceration, oedema (dropsy), peritonitis (parasite infestation) and nodules in the muscles that may cause deformation of the fish. The most susceptible fish are black mollies, neons, gouramis, and other carps, anabantids and tetras.

7. Vibriosis

It reveals multifocal necrosis and haemorrhage in visceral organs, mostly in liver. Necrosis and haemorrhage are also seen in the skeletal musculature. Bacterial foci may or may not be present. Abdominal distension due to ascites is also seen.

8. Enteric Septicaemia of Catfish (ESC)

It is caused by *Edwardsiella ictaluri* bacterium, which is a short, Gram-negative, pleomorphic rod with flagella. It mainly infects catfish and others like knife fish and barb. The disease can be either acute septicaemia or chronic encephalitis. Outbreaks normally occur in spring and autumn. In acute ESC, an acute septicaemia develops which shows multiple petechial haemorrhages that develop into depigmented ulcers. Other clinical signs are abnormal behaviour, exophthalmos, haemorrhagic gastroenteritis, oedema and ascites. Signs of chronic encephalitis or chronic ESC are abnormal behaviour, abnormal swimming patterns, swelling and ulceration of head and death. Any fish that survive the infection become latent carriers of disease.

9. Cotton Mouth

In fact, this condition is caused by the bacterium *Chondrococcus columnaris* that is popularly called *'mouth fungus'*, which is so named because it looks like a fungus attack of the mouth. It shows up first as a gray or white line around the lips and later as short tufts sprouting from the mouth like fungus. The toxins produced and the inability to eat will be fatal unless treated at an early stage. Symptoms are white cottony patches around the mouth and loss of appetite.

10. Furunculosis

In this, focal localization of bacteria in the dermis, gills, kidney, spleen, heart, liver or most visceral organs with little host response in some foci are the most common finding, but haemorrhage, oedema, macrophage infiltration and liquefactive necrosis may also be seen with some bacterial foci. Atypical furunculosis shows post-smolts with dermal haemorrhage and liquefactive haemorrhagic lesions in salmon parr muscle (Figure 41).

11. Bacterial Gill Disease (BGD)

This is primarily an epithelial disease with bacteria, colonizing the tips of secondary lamellae, spreading inwards and then resulting in a proliferative bronchitis, which causes epithelial hyperplasia. Fusion of secondary lamellae may occur distally enclosing the bacteria, sloughed epithelial cells and mucus, or may obliterate any lamellar space. The fusion of primary lamellae will also occur in the severe long-standing cases. The affected salmon fry fish shows mottled and swollen gills (Figure 42).

12. Bacterial Kidney Disease (BKD)

In BKD (Figure 43), chronic granulomatosis, mainly of haematopoietic tissue occurs, but extends to liver, cardiac and skeletal muscle or any organ. Granuloma is often large, with a central caseous zone bounded by epithelioid cells and infiltrating lymphoid cells. In aggressive infections, capsule is variable and encapsulation is lacking.

In the bacterial disease, *'pseudotuberculosis'*, the necrotized part in the centre is surrounded by the epithelioid matter. In another bacterial disease, *'red fin disease'*, extreme hyperaemia and haemorrhage in the interstitial region and endothelial tissue

Figure 41: Furunculosis in Salmon.

Figure 42: Mottled and Swollen Gills in BGD.

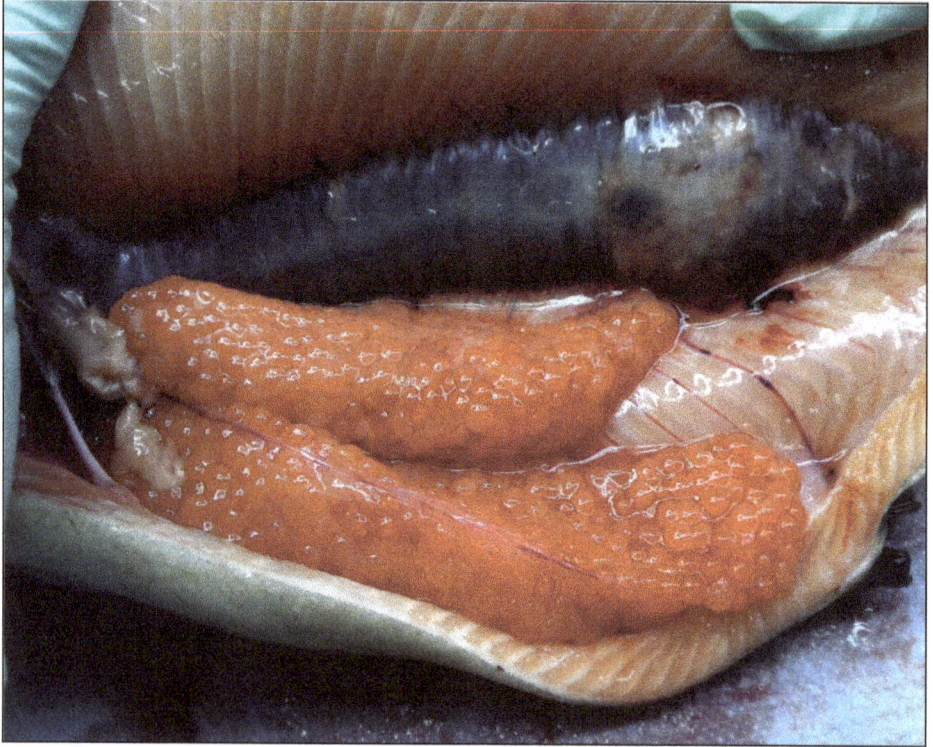

Figure 43: Bacterial Kidney Disease.

Figure 44: Goldfish Ulcer Disease.

of the atrium are noticed. In the *'red spot disease'*, haemorrhage is produced in the lower layer of epidermis, dermis and scales. *'Goldfish ulcer disease'* is caused by *Aeromonas salmonicida* bacteria in goldfish (Figure 44).

For fin and tail rot fin disease, one minute dip in 1:3,000 solution of copper sulphate ($CuSO_4$) should be given. Painting the site of infection with concentrated $CuSO_4$ solution also helps to cure the disease. In ulcer disease, badly infected fish should be destroyed and the pond water should be disinfected with a 0.5 ppm solution of $KMnO_4$. Dip treatment for one minute in 1:2,000 solutions of $CuSO_4$ for 3 to 4 days in fish showing early stage of infection is required. Addition of chloromycetin at 5 to 10 ppm to pond water and/or addition of oxytetracycyline to feed at 75 mg/kg of fish/day is done. In dropsy, thorough disinfection is done with 1 ppm of $KMnO_4$ solution, or dip treatment in 5 ppm of the same chemical for 2 minutes can be given. In eye disease during the initial stage of infection, chloromycetin (8-10 mg/L) bath for one hour should be given for 2 to 3 days. Disinfection of pond with 1 ppm $KMnO_4$, improvement of sanitary conditions and resorting to artificial feeding to fishes (in planktons is not sufficient) should be done (Madhuri and Pandey, 2014).

Common Fungal Diseases of Fish and their Treatment

The most common fungal diseases of fish are as under (Madhuri and Pandey, 2014; Pandey *et al.*, 2015):

1. Ichthyophonus or Swinging Disease

This is caused by the fungus, *Icthyophonus hoferi*. If a fish appears lethargic and exhibits a loss of equilibrium, it may have *'Ichthyophonus'* fungus disease. Although *Ichthyophonus* fungi are normally considered to cause marine fish, they do occur in freshwater fish from time to time. So, these fungi grow in fresh and salt waters, in wild and cultured fish, but are restricted to cool temperatures (36-68°F). The disease is spread by fungal cysts, which are released in the faeces and by cannibalism of infected fish. Because the primary route of transmission is through the ingestion of infective spores, the fish with a mild to moderate infection will show no external signs of disease. In severe cases, skin may have a *'sandpaper texture'* caused by infection under the skin and in muscle tissues. The infected fishes become lethargic, and if the brain is infected, they may exhibit a loss of equilibrium as well as staggered movements. Some fish may show curvature of the spine. Internally, the organs may be swollen with white to grey-white sores. Diseased fish shows curious swinging movements, so the disease is called *'swinging disease'*. Along with liver, particularly severely affected organs are spleen, kidney, brain, gonads and gills (all in salmonids), heart (in herring), and musculature and nerve tissues behind the eyes (both in sea fish).

2. Cottony Growth on Fish Body

The fungi are always present in aquarium which can become infectious when fish already have damaged gills or skin. Fungal infections are typically secondary infections. *'Body fungi'* will attack on the skin and cause ulcers or death in extreme cases. Large amounts of organic materials like decaying food or fish waste can cause a breakout of fungus in the aquarium. Open wounds can develop fungus. White or grayish patches appear on the skin or gills which look like cottony or woolly growth on the body of fish.

3. Saprolegniasis

It is caused by *Saprolegnia parasitica*, which often infects fertilized eggs in hatching *'hapas'*. Initially, this fungus attacks dead eggs and then spreads on to surrounding viable eggs, resulting in their spoilage as well. Lesions are focal, grey-white patches on skin or gills which when examined under water have a cotton-wool-like appearance (Figure 45), where hyphal filaments extend out into the water. Early lesions are almost circular and grow by radial extension around the periphery until lesions merge. At this later stage, the patches are often grey or brown in colour as mud or silt becomes trapped by the mycelium. Gills, mouth or bronchial cavity can also be affected. Internal infections in peritoneum or gastrointestinal tract (GIT) in younger fry can also be seen and results in high mortalities. Freshwater fish eggs are also very prone to infection. The fungus usually establishes itself focally, invading the dermis and extending laterally over the epidermis eroding it as it spreads.

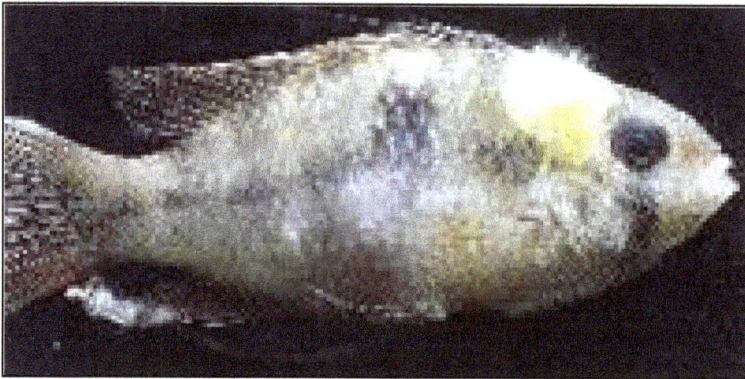

Figure 45: Saprolegniasis in Fish.

4. Epizootic Ulcerative Syndrome (EUS)

This disease is also called *'red spot disease (RSD), mycotic granulomatosis (MG) or ulcerative mycosis (UM)'*. It is caused by an oomycete, *Aphanomyces invadens* or *A. piscicida*. EUS is an epizootic condition of wild and farmed freshwater and estuarine fish. It is characterized histologically by penetrating hyphae surrounded by granulomatous inflammation. The fungal hyphae are seen growing in muscle as the lesion progress from a mild chronic active dermatitis to a severe locally extensive necrotizing granulomatous dermatitis with severe floccular degeneration of the muscle. Granulomas are formed around the penetrating hyphae. Red spots may be seen on the body surface, head, operculum or caudal peduncle (Figure 46). Large, red or grey shallow ulcers, often with a black-brown necrosis are noticed in the later stages. Most species other than striped snakeheads and mullet will die at this stage. In species like snakehead, the lesions become more extensive and can lead to complete erosion of the posterior part of the body or to necrosis of both soft and hard tissues of the cranium. Early EUS lesions are caused by erythematous dermatitis with no obvious oomycete involvement.

Figure 46: EUS Red Spots over Fish Body.

5. Branchiomycosis or Gill Rot Disease

It is caused by *Branchiomyces sanguinis* (in carp) and *B. demigrans* (in pike and tench) fungi. Branchiomycosis is a pervasive problem in Europe, but has been only occasionally reported by US fish farms. Both these species of fungi are found in fish suffering from an environmental stress like low pH (5.8-6.5), low dissolved oxygen or high algal bloom. *Branchiomyces* species grow at temperatures between 57°F and 95°F, but grow best between 77°F and 90°F. The main sources of infection are the fungal spores carried in the water and detritus on pond bottoms. *B. sanguinis* and *B.*

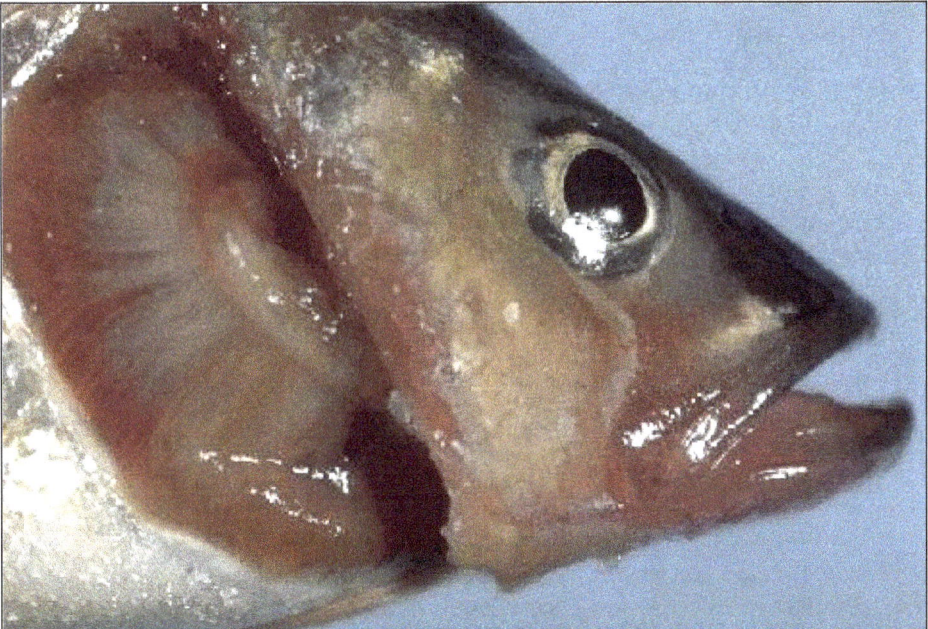

Figure 47: Branchiomycosis in Carp.

demigrans infect the gill tissues of fish. Fish may appear lethargic and may be seen gulping air at the water surface (or piping). Gills appear striated or marbled with the pale areas representing infected and dying tissue (Figure 47). Gills should be examined microscopically by a trained diagnostician for verification of disease. Damaged gill tissues with fungal hyphae and spores will be present. As the tissues die and fall off, the spores are released into the water and transmitted to other fish. High mortality is often associated with this disease.

Saprolegniasis is best prevented by good management practices like good water quality and circulation, avoidance of crowding to minimize injury (especially during spawning), and good nutrition. Once *Saprolegnia* is identified in an aquatic system, sanitation should be evaluated and corrected. Common treatments include $KMnO_4$, formalin and povidone iodine solutions. Over treatment can further damage fish tissues, resulting in recurring infections. Environmental management is essential for satisfactory resolution of chronic problems. Bath treatment in sodium hydroxide (NaOH, 10-25 g/L for 10-20 minutes), $KMnO_4$ (1 g in 100 L of water for 30-90 minutes), or $CuSO_4$ (5-10 g in 100 L of water for 10-30 minutes) should be given (Madhuri and Pandey, 2014).

Avoidance is the best control for branchiomycosis. Good management practices will create environmental conditions unacceptable for growth of fungi. If this disease is present, do not transport the infected fish. Great care must be taken to prevent movement of the disease to non-infected areas. Formalin and $CuSO_4$ have been used to help stop mortalities; however, all tanks, raceways and aquaria must be disinfected and dried. Ponds should be dried and treated with quicklime (calcium oxide). A long term bath in acriflavine neutral or Forma-Green for 7 days helps this condition. Ponds should be dried and treated with quicklime (calcium oxide) and $CuSO_4$ (2-3 kg/ha). Dead fish should be buried. There is no cure for fish with *I. hoferi* infection; they will carry the infection for life. Prevention is the only control. To avoid introduction of infective spores, never feed raw fish or raw fish products to cultured fish. Cooking helps destroy the infective life stage. If icthyophonus disease is identified by a trained diagnostician, it is important to remove and destroy any fish with disease. Complete disinfection of tanks, raceways, or aquaria is encouraged. Ponds with dirt or gravel bottoms need months of drying to completely eliminate this fungus (Madhuri and Pandey, 2014).

Common Protozoan Diseases of Fish and their Treatment

The most common protozoan diseases of fish are as under (Madhuri and Pandey, 2014; Pandey *et al.*, 2015):

1. Costiasis

This disease is caused by the mastigophoran (protozoan) parasite, named *Costia necatrix* (costia) or *Ichthyobodo necator*. Costiasis is the most common mastigophoran infection found in the Indian major carp. The lesions include the presence of a bluish coating on the skin and the presence of large amount of mucus. The parasite causes irritation and disturbs respiration. Other mastigophoran parasites recorded are *Bodomonas rebae* in mrigal, rohu and catla and *Trypanosoma puctati* in *C. punctatus*.

2. Ick Disease or Ichthyophthiriasis

This disease is a very contagious disease, caused by a protozoan ciliate, *Ichthyophthirius multifiliis* (freshwater ich), which infects different regions of the external body of major carps. Ich (ich) infestation is a very common pest among hobby aquarists. It can be introduced by rapid changes in water chemistry like temperature or pH (these weaken the immune system of fish, leaving them susceptible to disease), poor water quality, or bullied fish. This parasite causes simple hyperplasia of the epidermal cells around the site of infection, causing formation of pustules or small, grain-like white spots. Thus, this disease is also called *'white spot disease (WSD) or ich'*. The parasite spends a part of its life within the skin, gill and fins of fish, where they form thickened masses of protective mucus as *'white spots'* (trophonts or cysts). Ick lesions appear as small white cysts (*sprinkles of salt*) on the skin, gills and fins (Figure 48). Fish may try to rub itself against decor or hard surfaces in aquarium, giving an impression of scratching an itch. Once the cysts mature, they fall down to substrate and give rise to thousands of new free-swimming parasites which are free to prey upon other fishes. Other symptoms are loss of appetite, rapid breathing (rapid gill fluctuation), cloudy eyes, fin deterioration, flashing, exhaustion and isolation (fish may hang out near the surface or water return). Ichthyophthiriasis has been experimentally produced in the fingerlings of *Labeo bata* and *C. mrigala*. An outbreak of ick is very difficult to control if not noticed in earlier stages. Since the parasite is released into aquarium, the entire aquarium has to be treated.

Figure 48: Ick Disease in Fish

3. Trichodiniasis

This disease (Figure 49) is caused by a group of peritrichal ciliated protozoans. The organisms are saucer shaped, 50 μ in diameter, with rows of cilia at both ends and a macro- and micro- nucleus. When viewed dorso-ventrally, the parasite appears as an ornate disk with a characteristic ring of interlocking denticles, forming a circle in the middle of organism. *Trichodina truttae* is a specific pathogen for salmonids. These are seen on the most fresh and salt water fishes. This protozoon is relatively common on many fishes, and is not always associated with the disease. Clinically, the affected fish usually exhibits the flashing and becomes lethargic. There is increased

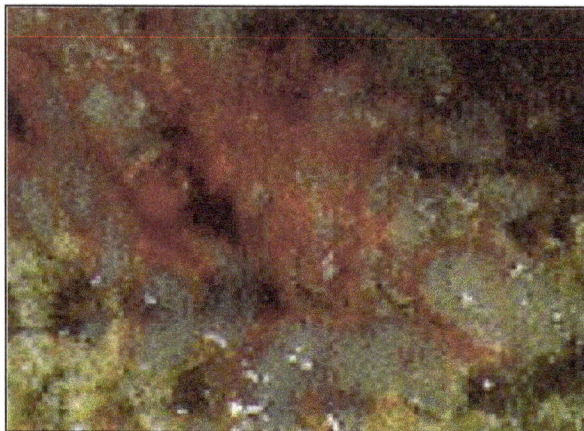

Figure 49: Trichodiniasis.

mucus production, causing a white to bluish haze on the skin. The skin may develop ulcers and the fins may fray. If gills are involved, the fish may have severe respiratory distress. Histologically, masses of organisms are attached by adhesive discs and denticles of exoskeleton to the epidermis. The underlying epithelial cells undergo necrosis. There is secondary hyperplasia and hypertrophy of the gill epithelium.

4. Amoebic Gill Disease (AGD)

This disease is caused by a protozoan parasite, *Neoparamoeba perurans*, which is the most important amoeba in cultured fish. Certain bacteria and fungi present in the water can also cause gill disease. AGD is a potentially fatal disease of some marine fish. It primarily affects farm raised fish of the Salmonidae family, most notably affecting the salmon (*Salmo salar*). Turbot, bass, sea urchin and crab have also been infected. Symptoms typically begin to appear two months after the fish are transferred from freshwater hatcheries to open net sea cages. In this disease, the fish appear to be swimming in a great hurry, but it does not move at all. The fins flap very fast and fish gasp for breath at the surface, or lie listless at the bottom. The gills become swollen and discoloured. Mucus is built-up on the gills of infected fish and hyperplastic lesions appear, causing white spots and eventual deterioration of gill tissue (Figure 50). Fish will show signs of dyspnoea such as rapid opercular movements and lethargy. Histopathologically, thickening of gill lamellae as pale areas is observed. Although recognized by hyperplastic and proliferative gill lesions, the effect of AGD occurs before transfer of oxygen across the gill is severely compromised. AGD affected fish show a significant increase in vascular resistance contributing to cardiovascular collapse. Such effects result in compensatory changes in heart shape to improve its efficiency at pumping blood.

5. Hole in the Head Disease or Hole in Head

It is caused by *Hexamita* (*H. salmonis* and *H. truttae*) parasite. It usually infects large cichlids like discus, pike and Oscar fishes. *Hexamita* is a small parasite of about the same size as a RBC. It is oval in shape, with two nuclei, six flagella on the front of

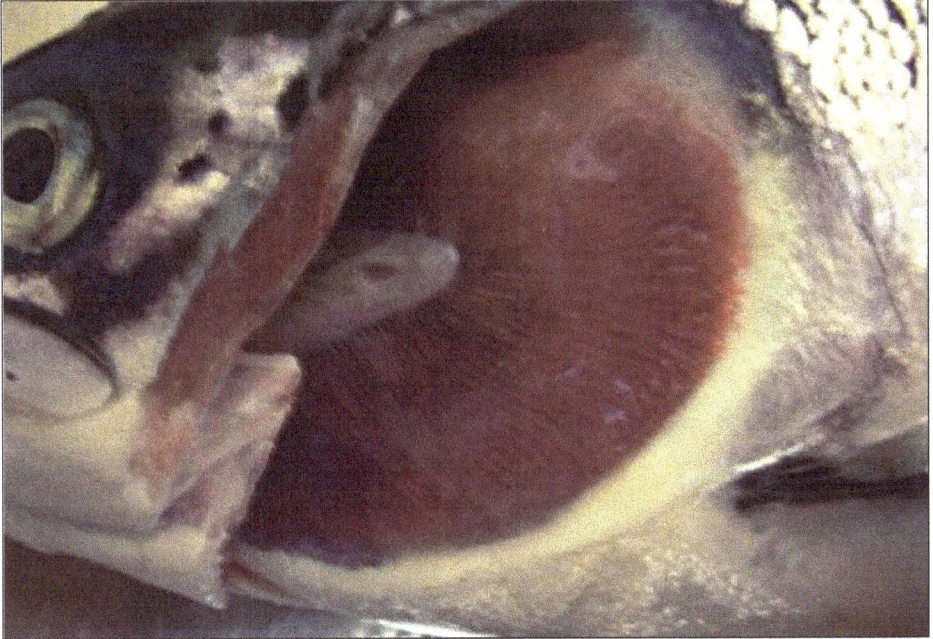

Figure 50: Amoebic Gill Disease in Salmon.

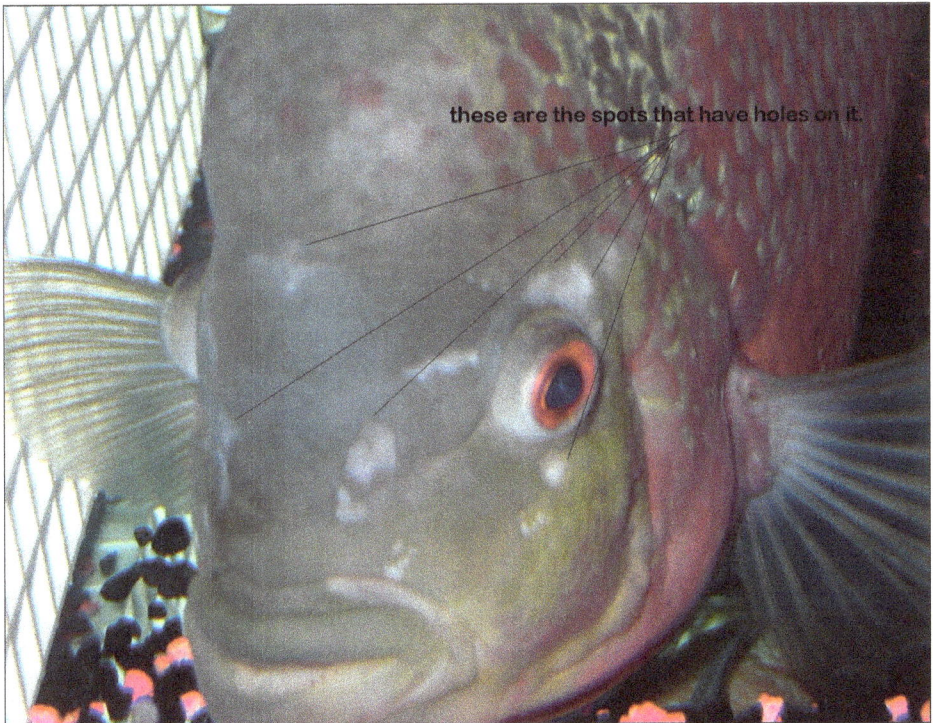

Figure 51: Hole in the Head Disease.

body and two flagella at the rear end. The infection is usually a secondary infection. Low levels of this parasite live within the fish, and poor water quality and/or poor diet may lead to infection. *Hexamita* initially infects intestines, but can rapidly spread to liver and blood. As the name of disease suggests, the infected fishes develop small pits, mainly in the head region. *Hexamita* can build up under skin around the head of infected fish, which can cause tissue breakdown of localized areas, leading to *'hole-in-the-head disease'*. The holes are usually white in colour (Figure 51), and have yellowish mucus trailing from them. Infected fishes become listless, lose colour and stop feeding (loss of appetite). Other symptoms are ulceration on body, pale-coloured mucus like faeces and emaciation. In untreated fishes, entire areas of skin may be undermined, resulting in large open sores. Hole in head is a very serious disease which warrants prompt treatment.

6. Velvet Disease

It is also called *'gold dust disease'* (Figure 52). It is caused by *Oodinium* (also called *Piscinoodinium*), a genus of microscopic protozoan dinoflagellates. These parasites live off salt and fresh water fishes. Velvet (*Oodinium*) is most often confused as ich parasite. The difference is, velvet is smaller and infects predominately to body and looks like a fine powder rather than salt sprinkles. Velvet parasite is classified as a parasitic alga, because it contains chlorophyll and, so obtains some of its food through chlorophyll. For this reason, it is often suggested to darken the tank if the fishes show a velvet outbreak, as chlorophyll requires

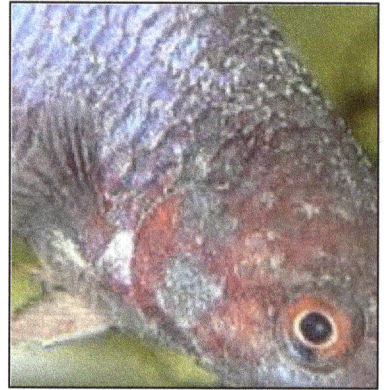

Figure 52: Velvet Disease.

visible light to survive. Velvet parasite has two life stages: a free-swimming form and a cyst form. Infective stage of this parasite is the *'free-swimming stage'*. During this stage, the velvet parasite has two flagella which enable it to propel itself through the water. It propels itself through the water until it finds a suitable host. Then, it will attach itself to skin or gills of fish. Eventually, the parasite will form a cyst on the fish, which will remain on-site until it releases several hundred free-swimming *Piscinoodinium*. These newly released *Piscinoodinium* go in search of another host, and the cycle begins again. For this reason, velvet is very contagious and mostly caused by stress, poor water quality or chilling (sudden changes in water temperature). Velvet attacks fish body and produces small yellowish spots, which are much smaller than the ich spots. Similar to fish infested with ich, the velvet-infested fish may show clamped fins, and they may flash off rocks and other surfaces in an attempt to dislodge the parasites. If gills are affected, the fish may exhibit rapid gill fluctuation (breathing heavily) or gasp for air at the surface. Due to disease, the affected fish will dart around and flash or scrape itself against aquarium decor. Fine yellowish or white dusts on the body and clamped fins are also seen. It is a bit easier to cure than ich, as its life cycle is shorter. However, this disease can prove fatal, if left untreated.

7. Neon Tetra or Pleistophora Disease

'Neon tetra disease' is so named because the neon tetra (*Paracheirodon innesi*, a freshwater fish of characin family) is generally infected by the causative agent, *Pleistophora hyphessobryconis*. It is a single-celled protozoan parasite, called a *'sporozoan'* (the cell, releases millions of spores into water during reproduction). Although this parasite has been reported to infect the most characins, cichlids, cyprinids (danios) and all carp-like fishes are also infected. It typically enters the body of host fish through contaminated food. The parasites travel through the fish's blood, quickly spreading to all areas of body. They not only affect the fish's muscles but all other internal organs. Once the parasite has entered the intestine of fish, it spreads and eventually settles in the skeletal muscles where it begins to mature. The spores grow and replace the muscle fibers, and cause degeneration and wasting of muscle. Loss of colouration and pale, white patches under the skin, particularly strip of skin beneath the dorsal fin are seen (Figure 53). Other symptoms may be spinal curvature, emaciation, fin deterioration and erratic swimming behaviour. Over time, this parasite produces spores which may be distributed to new host fish upon the death of original host (infected fish). If parasite settles in the gut or kidneys rather than the skeletal muscles, it may release spores directly into aquarium through eliminated waste. Because there are a variety of ways for the spores to be transmitted, the pleistophora (neon) disease is highly contagious.

Figure 53: Neon Tetra Disease.

8. Swim Bladder Disease (SBD)

It is also known as *'swim bladder disorder or flip over'*. This is a common ailment in the aquarium fish, and is caused by intestinal parasites, mainly protozoans and helminthes. It can also be produced by deformed bladder or bacterial infections, or by constipation induced by high nitrate levels from overfeeding. SBD mostly affects the

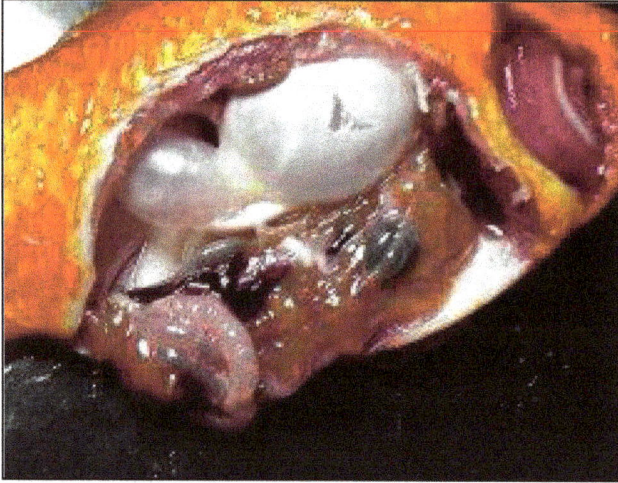

Figure 54: Swim Bladder Disease.

fishes that have globoid body shape, and the fancy goldfish is most commonly affected. The swim bladder is an internal gas-filled organ which contributes to the ability of a fish to control its buoyancy, and so to stay at the current water depth without having to waste energy in swimming. Therefore, any damage to buoyancy (a membrane) can cause SBD, and loss of buoyancy is one of the first indications of this disease (Figure 54). The affected fishes can float nose down tail up, or can float to the top (float at the surface only) or sink to the bottom of the aquarium (stay at the bottom of the aquarium) and unable to rise. Thus, the loss of balance and/or abnormal swimming in the affected fish is observed.

9. Epistylis or Red Sore Disease

It is caused by branched stalked ciliated protozoan, *Heteropolaria colisarum*. It is found primarily in wild populations of scaled fish. This protozoan has also been observed on eggs. Clinically, one sees ulcers or cotton-like growth on the skin, scales and spine, resulting in a red coloured lesion. In catfish, the lesions involve the spines and bones that underlie the skin of head and pectoral girdle.

10. Myxosporidians Infestation

Myxosporidians constitute typical fish parasites known to produce cysts on different regions of the body and internal tissues and organs. The common myxosporidians genera are *Leptotheca, Chloromyxum, Myxobolus, Henneguya, Thelohanellus, Myxidium, Lentosperma*, etc. Symptoms of myxosporidians infestation include weakness, emaciation, raising of scales along their posterior margins and falling of scales, etc.

For ich disease, hourly dip, in 1:5,000 formalin solutions for 7 to 10 days, or in 2 per cent common salt solution for more than 7 days, or in 1:50,000 quinine solutions for 3 to 10 days should be given. Control measure for costiasis disease is the bath of 10 minutes in 3 per cent common salt solution. *Bodomonas rebae* (which causes costiasis) is killed in 5 to 10 minutes by 2 to 3 per cent common salt solution. In trichodiniasis, chelated Cu compounds, which are extremely effective against protozoan parasite, is used. Cu compounds, *e.g.*, argant and aquavet are available in market (Madhuri and Pandey, 2014).

Common Crustacean Diseases of Fish and their Treatment

The most common crustacean diseases of fish are as under (Madhuri and Pandey, 2014; Pandey *et al.*, 2015):

1. Argulus or Fish Lice Infestation

Argulus parasites are very major threat to fish health. They are small, white lice apparent on the fish body, and fish rubs against hard surfaces in aquarium. These crustaceans can cause significant morbidity and mortality when they heavily infest to fish. They are also vehicle for other fish diseases. The affected fish shows patches of haemorrhagic and oedematous skin, gills or fins (Figure 55). *Argulus* causes these injuries by attaching to fish with its curved hooks and sucker. Its feeding apparatus further injures host fish, when it inserts stylet into epidermis and underlying host tissue, causing haemorrhage. *Argulus* parasites feed on the host's blood and body fluids. Feeding apparatus also releases digestive enzymes, which cause systemic illness. *L. rohita* is mainly affected by this disease.

Figure 55: Argulus Infestation.

2. Learnaea Disease

It is caused by *Laernaea cyprinacea*. Its body is elongated, worm like and its head is embedded in the body of fish. Many branches arise from this embedded head. Due to infestation, initially fish starts swimming rapidly and later on, lesions/wounds develop on the body. Catla fish is mainly affected by this disease.

An isopod parasite, *Ichthyoxenus jellinghausii,* has been seen in *L. bata* and *L. gonius* of Pariat lake, Jabalpur (MP). These parasites do not cause apparent harm to hosts.

For *Argulus* infestation, pond treatment with gammexane at a concentration of 0.2 ppm should be repeated at weekly intervals twice or thrice. In learnaea disease, when a few fishes are affected, mechanical removal of parasites by pulling them out from anchorage by means of fine forceps would be sufficient, preferably followed by a bath in weak $KMnO_4$ solution for 2 to 3 minutes. In case of large numbers of fish being infected, dip treatment over short durations in a 5 ppm solution of $KMnO_4$ has been suggested (Madhuri and Pandey, 2014).

Common Helminth Diseases of Fish and their Treatment

Many parasitic worms infect fish and cause great harm to them. *Dactylogyrus* and *Gyrodactylus* are the two common parasitic worms. *Dactylogyrus* attacks the gills, while gills and skin both are attacked by *Gyrodactylus,* causing gyrodactylosis disease (Figure 56). The fish becomes less motile by the attack of these parasites, their fins start falling, body becomes yellow in colour and the blood spots develop on their body. Delay in treatment may be lethal. Cyst of the metacercaria of the worm, *Posthodiplostomum cutiola* also causes 'black spot disease' in fish. Such black spots appear on the whole body, including eyes and mouth. These parasites enter in the body of the fish through their skin during cercaria stage. 'Skin and gill fluke disease' is caused by the flukes, which attack the gills and skin of infected fish. Small numbers of flukes are pretty much harmless, but in bad water conditions, overcrowding, or aquaria with high organic content, they can multiply at a fast rate. The fish will flash or scrape itself against aquarium decor, have rapid gill movement, and at a more advanced stage, the fish will isolate itself and lie at the bottom with fins clamped; cloudy skin is also seen due to excess mucus (Madhuri and Pandey, 2014; Pandey *et al.,* 2015).

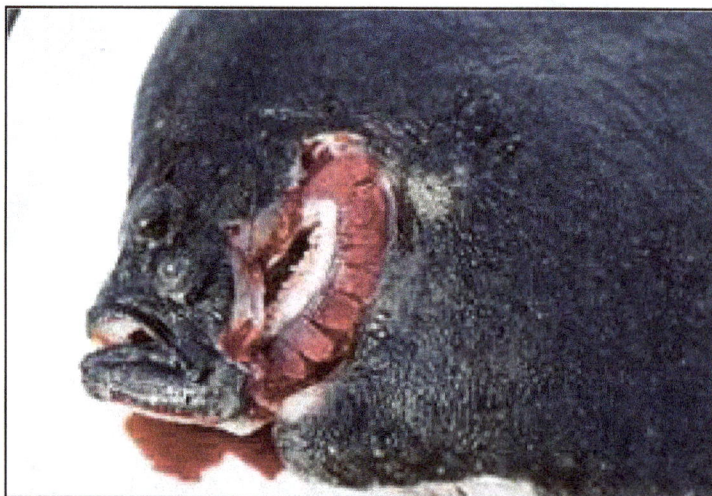

Figure 56: Gyrodactylosis on Skin of Atlantic Salmon.

For gyrodactylosis (caused by *Gyrodactylus*), alternate bath in 1:2,000 acetic acid and sodium chloride (NaCl, common salt) solutions have been effective for major carp. Dip treatment of 5 minutes duration in 5 per cent NaCl, or of 5 to 10 minutes duration in 1:5,000 formalin solution should be given (Madhuri and Pandey, 2014).

Nematodal Diseases of Fish

The nematodes are generally present even in the healthy fishes, but they can turn into a problem when a fish becomes weakened due to some other reasons. In the nematodal infestation, the anus of fish may protrude a little, and a thin threadlike worm will drift out of the anus. The infected fish becomes listless and bloated, and refuses to feed (Pandey *et al.*, 2015).

Drugs Against Parasitic Diseases of Fish

Certain Parasitoses in Fish

'Parasitosis' (parasitic infestation/infection) in fish is very common. The intestinal infections of fish by many parasites cause severe morbidity and mortality. For example, the flagellates of *Hexamita* parasite are often associated with high mortality in fish. The pathological effects arise when the host is weakened by other factors like inadequate diet, change in diet, low oxygen content in the water, overcrowding, inappropriate handling and/or keeping fish of different sizes together. The flagellate parasites interfere with nutrition by competing for essential nutrients and/or by damaging intestinal epithelium. In both salmonids and tropical aquarium fish, the intestinal parasitosis is often pathogenic only when the number of parasites present is very high (Pandey, 2013a).

'Hexamitosis' (a disease caused by *Hexamita* parasite) is probably the most frequent internal flagellate parasitoses of fish, notably in the young salmonids, though also in carp, aquarium species and various marine fish. Heavily infected fish are weak, listless, anorexic and emaciated, so that the head of a particular fish appears large with respect to the body (*'pinhead fish'*). The affected fish typically swim on their side, or with corkscrew movements. The fish population affected by the acute hexamitosis shows high mortality over a very short period, due to rapid multiplication of the parasite and associated damage to the intestinal epithelium. Chronic form of hexamitosis is likewise common, and generally occurs between spring and autumn; mortality per unit time is only slightly higher than in healthy fish, but severe losses may occur because the situation continues for a period of weeks (Pandey, 2013a).

In trout and other salmonid fish infected with *H. salmonis* parasite, the adverse effects commonly observed are anaemia, weight loss, dark colouration, enteritis, excessive body mucus and yellowish intestinal mucus (attributable to modified release of bile into the digestive tract). In addition, the intestinal haemorrhage and liver cell necrosis may also be observed. It is worth noting that the *H. salmonis* infected fish passes both trophozoites and cysts in faeces. The fish that appear to be heavily infected (*i.e.*, numerous parasites in the pyloric caecae and the intestine) may show no signs of damage to the mucosa, and no evidence of invasion of the epithelium by the parasite. Cysts were not observed at any stage, even after subjecting samples to concentration methods, *e.g.*, Bailinger method. Cysts thus appear to be very rare (Pandey, 2013a).

Some Antiparasitic Drugs Against Parasitic Diseases of Fish

Many drugs have been investigated for the treatment of parasite infection in different species of fish. However, the only oral pharmacological treatment of *H. salmonis* determined to date have been the drugs of nitroimidazole group, which show the activity against different protozoan groups, including flagellates and ciliates. Hence, the current treatment of choice is dimetridazole or metronidazole in the feed. Further, there have been few studies of the possible anti-*Hexamita* or antiparasitic effect of the drugs of other groups. Treatment for ectoparasitic diseases in freshwater fish with formalin seems at present to be ineffective. Formalin possibly leaves toxic residues in fish flesh and in the environment which are eventually harmful to the consumers. The alternative way to solve this problem is to use traditional medicinal plants instead (Chitmanat *et al.*, 2005).

Besides above, there are many other pharmacological treatments of fish parasitic diseases (Madhuri and Pandey, 2014; Pandey, 2013a), as illustrated below:

The results a study confirmed the efficacy of nitroimidazoles against the parasitic diseases of fish. In this study, *H. salmonis* infection in the rainbow trout (*O. mykiss*) fish was completely eradicated not only by the metronidazole (which has been recommended earlier for the treatment of hexamitosis), but also by benznidazole, ronidazole and secnidazole). The non-nitroimidazoles, *e.g.*, albendazole, amnosidine, diethethylcarbamazine and nitroscanate also completely eliminated the infection. The remaining non-nitroimidazoles tested (amprolium, bithionol, febantel, flubendazole, levamisole, netobimin, niclosamide, nitroxynil, oxibendazole, parbendazole, piperazine, praziquentel, tetramisole, thiophanate, toltrazuril, trichlorfon and triclabendazole) were found not effective. The four nitroimidazoles (albendazole, aminosidine, diethylcarbamazine and nitroscanate) completely eliminated the infection at the dose of 5 g per kg feed for 2 days. All nitroimidazoles except metronidazole were effective even at the lower dose of 2 g per kg feed for 2 days. These results confirmed the efficacy of these drugs when administered for a shorter period and at much lower doses than the other drugs tested. The only non-nitroimidazole drugs that completely eliminated infection were albendazole, aminosidine, diethylcarbamazine and nitroscanate. Out of albendazole, aminosidine, diethylcarbamazine and nitroscanate antiparasitic drugs, the only one previously recommended for the treatment of infection by *H. salmonis* is aminosidine (15 g/kg feed for 3 consecutive days). Nitroscanate appears to have a rather broad activity spectrum, since it has been shown to be effective for bath treatment of *Gyrodactylus*. Neither albendazole nor diethylcarbamazine has previously been shown to be effective for the treatment of protozoan parasitosis of fish. None of the antiparasitic drugs, *viz.*, albendazole, aminosidine, diethylcarbamazine and nitroscanate showed negative effects (signs of toxicity and behavioural effects, including anomalous swimming movements, rejection of food), suggesting that all the above four nitroimidazoles are viable options for the treatment of infection of salmonids by *H. salmonis* parasite. This is of particular interest because *H. salmonis* strains apparently resistant to metronidazole have appeared in some farms.

Three drugs have also been recommended for the treatment of protozoan parasitoses of fish, *viz.*, amprolium, bithionol and toltrazuril. Of these, toltrazuril has been most widely used for the treatment of fish parasitoses, and indeed this drug has been recommended for the treatment of various microsporidian and myxosporidian infections. This drug is not, however, effective for the treatment of infestation by *Ichthyophthirius multifilils* ectoparasite ciliate. Among amprolium, bithionol and toltrazuril, the amprolium was found to be effective after oral administration; indeed, its administration by oral route is recommended for myxosporidiosis. Bithionol and toltrazuril have been shown to be effective only in bath treatments. In a study of efficacy for the treatment of infestation by the *Ichthyobodo necator* flagellate, complete elimination of infestation in all fish assayed was achieved after bathing with bithionol at 25 mg l^{-1} for 3 hr on 2 consecutive days, but not after bathing with amprolium or toltrazuril.

Nitroimidazoles are currently the only drugs recommended for oral treatment of *H. salmonis* infection. Earlier reports have recommended dimetndazole at the dose of 1.5 g per kg feed for 3 days, or at 15 g per kg feed for 4 to 7 days. Metronidazole has been recommended @ 0.5 mg per kg feed for 2 days, 20 mg per kg feed for 2 days, or 1.5 g per kg feed for 3 days. Bath treatment with metronidazole has been recommended for infections with *Trichodina, Ambiphyra and Chilodonella.*

Potassium Permanganate Against Infections of Fish

Potassium Permanganate and its Remedial Importance in Fish

$KMnO_4$ is an oxidizing agent which has been used for many years in the aquaculture, can be used to *'disinfect'* the external surface of fish. It is also used in water conditioning systems and in the plumbing industry. As an oxidizer, it is able to chemically *'burn up'* the organic materials. The undesirable organic matters like bacteria, parasites and fungi, as well as desirable materials like gill tissue and mucus can be destroyed by $KMnO_4$. Because $KMnO_4$ cannot distinguish between desirable and undesirable organic matter, it is up to the individual to use $KMnO_4$ in a manner that results in maximum benefit and minimum harm to the treated fish. When $KMnO_4$ is active (in its unoxidized form), the treated water turns a pinkish-purple colour. As the chemical is *'deactivated'* (by oxidizing organic material), the water colour changes to yellow or muddy brown. This colour change is an important tool when monitoring the chemical treatment; however, this may make $KMnO_4$ undesirable for the use in display tanks, exhibits, or ornamental ponds. As with many chemicals used in the water, $KMnO_4$ is harmful to the plants and invertebrates (Pandey, 2013b).

External infection in fish, including ornamental fish is a great problem. If uncontrolled, the infections may become systemic, leading to death of affected fish. Thus, potassium permanganate ($KMnO_4$) can effectively control several bacterial, parasitic and fungal infections. However, this chemical is not viricidal. If $KMnO_4$ is not used in proper time, the systemic infections may develop which requires antibiotics and other specific remedies. Thus, by using $KMnO_4$ in early stage of infection, the fish owner can save money as the use of expensive antibiotics is decreased, thereby decreasing the incidence of resistant bacterial strains. In the ornamental trade, proper use of antibiotics at the onset of an infection can also speed the movement of fish as they do not need to be held for lengthy (usually 10-14 days) antibiotic treatment (Pandey, 2013b and 2014a). $KMnO_4$ is reasonably safe to use in recirculation systems, and has minimal impact on the biofilters when used at concentrations of 2 mg/L. Treated water should retain purple colouration for at least 4 hr. There is extensive information on the use of $KMnO_4$ in freshwater systems, but much less is known about its effect in marine systems (Pandey, 2013b).

Treatment Dosage of Potassium Permanganate

For most fish, $KMnO_4$ can be used at a concentration of 2 mg per L as a long-term bath (minimum 4 hr) in fresh or salt water system of fish. Fish culturists should run a small bioassay before treating the fish with $KMnO_4$. Some fish, including certain lake cichlids, are sensitive to $KMnO_4$ and lower concentrations (1 mg/L) may be safer. Fish farmer may know the species sensitivity by seeing the behaviour of fish during treatment. This is especially important when treating a species for the first time. If fish react adversely, immediate action (*e.g.*, diluting $KMnO_4$ with freshwater) should be taken. Because $KMnO_4$ is deactivated by organic matter, it may be necessary to increase the amount added to ponds or other systems where organic material has been allowed to accumulate. A safe way of doing this is to add $KMnO_4$ to the system in 2 mg per L increments. If water colour changes from purple to brown in less than 4 hr from the start of the first treatment, an additional 2 mg per L should be added. If a total application of $KMnO_4$ @ 6 mg per L does not result in maintenance of purple colour for at least 4 hr, the system should be cleaned. Most of the organisms that are treated with $KMnO_4$ thrive in an organically rich environment; so the improved sanitation can have a tremendous impact on treatment efficacy. $KMnO_4$ can also be used as a short-term bath @ 10 mg per L for 30 minutes. At this concentration, careful observation of fish is mandatory to avoid the fish mortality. This is a convenient treatment when the fishes are being removed from ponds and brought into the buildings for sorting and shipping. Following a $KMnO_4$ treatment with a low concentration (2-10 ppm) of NaCl solution as a semi-permanent treatment for many days or weeks (depending on the species treated), may be beneficial. This combination is especially effective in minimizing the *Columnaris* infection (columnaris disease) after handling the fish. $KMnO_4$ can be used as a surface disinfectant at concentrations of 10 mg per L (30-60 minutes contact time) to 500 mg per L (30 seconds contact time) in a fish room or hatchery; however, quaternary ammonium compounds are better for this purpose (Pandey, 2013b and 2014a).

Dose Calculation and Stock Solution of Potassium Permanganate

To calculate the dose of $KMnO_4$ required, the simple formula is:

Amount of $KMnO_4$ = volume × conversion factor × treatment rate

If the pond or tank volume is measured in gallons, the conversion factor is 0.0038. So for a treatment rate of 2 mg per L, the formula would be:

Quantity (g) of $KMnO_4$ = gallons treated × 0.0038 × 2 mg per L.

Hence, to treat a 250-gallon vat, the quantity (g) of $KMnO_4$ = (250 gallons) × (0.0038) × (2 mg/L) = 1.9 g. Generally, one level teaspoon of $KMnO_4$ weighs about 7 g.

An alternative method of measuring $KMnO_4$ is to mix a *'stock solution of KMnO_4'*. A *'stock solution'* is a concentrated solution of chemical from which small amounts can be taken to treat tanks as needed. This is useful when either multiple tanks or multiple treatments are needed. An easy way to make up a stock solution of $KMnO_4$, purchase a one-gallon bottle of distilled water, weigh 285 g of $KMnO_4$, add it to the solution and mix thoroughly. This stock solution will deliver a dose of 1 mg per L when delivered @ 1 drop per gallon. Thus, to obtain the desired concentration of 2 mg per L, the stock solution can be delivered @ 2 drops per gallon. The stock solution should be stored in a cool-dark area, and be replaced annually. When treating the larger systems, it should be noted that 20 drops are equal to 1 ml (or 1 cc), if measuring liquid with a syringe. Hence, 1 ml of stock solution will treat 10 gallons of water with a $KMnO_4$ concentration of 2 mg per L (Pandey, 2013b and 2014a).

Failure of Potassium Permanganate Treatment

Treatment failure of $KMnO_4$ is generally due to one of these three reasons:

(a) Excessive organic matter in the system resulting in rapid degradation of the $KMnO_4$;

(b) Incorrect or incomplete diagnosis; or

(c) Incorrect calculation or measurement of amount of $KMnO_4$ required. If treatment failure occurs, the sick fish should be taken to a diagnostic laboratory for an accurate diagnosis. Volume of the water treated, accuracy of calculations to determine treatment rate and accurate measurement (by weight) of $KMnO_4$ used are necessary for giving of an appropriate dose of $KMnO_4$. Since an excessive amount of organic matter in the system will result in rapid deactivation of $KMnO_4$, the contact time with active $KMnO_4$ will be inadequate for effective treatment. This is often a problem in heavily stocked ponds (Pandey, 2013b and 2014a).

Precautions in Using Potassium Permanganate

Although $KMnO_4$ is safe to handle but all chemicals should be handled with precaution. $KMnO_4$ will easily stain clothing and skin, so care should be taken during handling. Brown discolouration of skin is not painful, but it may be unsightly and takes many days to disappear. The brown stains to clothing can be permanent. The protective eye wear, gloves and clothing are recommended when handling the $KMnO_4$. The fish farmers and aquarists do occasionally mix the chemicals. It is important that $KMnO_4$ should *'never'* be mixed directly with formalin because the explosion or fire may result (Pandey, 2013b and 2014a).

As stated earlier, $KMnO_4$ is an indiscriminate oxidizer, so it can burn gill tissue and mucus of treated fish if too much quantity is used. A good rule of thumb to prevent excessive damage to fish is to avoid treating them with $KMnO_4$ more than once a week. If a chemical treatment is needed for a condition that requires more frequent application, *e.g.*, treatment for an outbreak of *'ich disease'* in fish, $KMnO_4$ is not a good choice (Pandey, 2013b and 2014a). When $KMnO_4$ is active (in its unoxidized form), the treated water turns a pinkish-purple colour. As the $KMnO_4$ is deactivated (by oxidizing organic material), the water colour changes to yellow or muddy brown. This colour change is an important tool when monitoring this treatment. However, it can make the $KMnO_4$ undesirable for use in display tanks, exhibits, or ornamental ponds. As with many chemicals used in water, $KMnO_4$ is harmful to plants and invertebrates. $KMnO_4$ solution in water must retain a purple colour for at least 4 hr, although some species may be sensitive to it and may not tolerate a full strength (2 mg/L) bath. Because of its harsh oxidizing properties, $KMnO_4$ should not be applied to fish more frequently than once per week, otherwise mortality may result. It is safe to use in marine and recirculation systems at the dose of 2 mg per L (Pandey, 2014a).

Disinfectants and Antibiotics for Ornamental Fish

Importance of Ornamental Fishes

'*Ornamental fishes*' are important commercial components of fisheries which provide for aesthetic requirements. Relatively minimum requirement of space or attention compared to any other pet animal is the reason for growing interest in keeping aquaria all over the world. Setting up of aquarium is also relatively less expensive, and can be fitted at any location of a small house (Pandey, 2013b and 2014b). So, keeping of ornamental fish as pet is becoming a popular hobby in the world, next only to photography. Millions of houses in the whole world have fish aquaria. Ornamental fish farming is also growing to meet this demand. In India, the West Bengal state plays pioneer role for production of ornamental fishes. Ornamental fish trade also helps, to some extent, in improving socio-economic condition of rural people and in upliftment of the condition of rural women in India. A rich diversity of fish fauna, low investment in the business, cheap labour, easy distribution channel and export centres help this business to improve most. However, instead of a high demand in foreign markets, this business is a '*neglected trade*' due to the huge losses faced by culturists by various disease outbreaks (Pandey, 2014b).

Different Ornamental Fishes and their Breeding

Ornamental fishes are grouped mainly into two categories: egg-layers ('*oviparous*') and live-bearers ('*viviparous*'; *e.g.*, guppies, platys, mollies and sword-tail). Some are categorized under egg-layers with no care (non-guarders), egg-layers with care (guarders), egg buriers (*e.g.*, kalli fish), mouth-incubators, nest-builders (*e.g.*, bettas and gouramies), egg-carriers, egg-scatterers (with adhesive eggs; *e.g.*, tetra, barbs and goldfish), egg-scatterers (with non-adhesive eggs; *e.g.*, zebra group), egg-depositors (*e.g.*, rasbora group) and live-bearers, etc. (Helfman *et al.*, 2009; Madhuri, 2013; MANAGE, 2008; Pandey, 2013b and 2014b).

'*Barbs*' are the most important egg-layers group. Most species of the barbs, *viz.*, rosy barb, tiger barb, striped barb and aruli barb, are of the Indian origin. Another main species of egg-layers group is '*danios*', which include giant danio, pearl danio and zebra danio. The zebra danio is the typical example of ornamental fish species of Indian origin which is bred and reared very easily. Among the '*rasbora*' species of egg-layers group, the slender rasbora, glowlight rasbora and scissor tail are the important ones. *'Goldfish'*, *C. auratus*, is the most commonly available fish, preferred by most of the hobbyists because of its attractive colouration ranging from pure gold to red, orange, black and albino. Different morphological characteristics such as divided caudal and anal fins of varying sizes, egg-shaped body due to shortened vertebrae, enlargement and protrusion of eyes, presence of pearl-like glittering scales, thickening of skin on the top of head, absence of dorsal fins, transparent and dazzling colours received through the cross breeding or selective breeding make the goldfish species more attractive. Some common varieties of goldfish are common goldfish, lion head, oranda (a modification of lion head), fringe tail, veil tail, fantail, shubunkin (scaleless hybrid), telescopic eye fish, etc. The goldfish grow up to 20 cm in length, but start breeding when they are only 6 cm long. *'Tetra egg-layers'* are small fishes of 3 to 8 cm long, majority of which have been originated from South America. The most common species of the tetra are black widow tetra, serape tetra, rosy tetra, lemon tetra, flame tetra, neon tetra, cardinal tetra, glowlight tetra, head and tail light tetra, pretty tetra, etc. *'Betta fish'*, *e.g.*, *Betta splendens* (commonly called *'Siamese fighting fish'*) occurs in varied colours like green, red, blue, albino and sometimes with a combination of 2 or 3 shades. Attractive colour and hardiness of this species are the characters for wide adoption by the hobbyists. The males of betta are brightly coloured with beautifully spread-over fins. They show aggressive behaviour only when other males are present. *'Angel fish'* species are widely preferred for aquaria, with varieties like black, veil tail, marble and albino. The filamentous lower fins and their compressed body shape with their elegant movements are pleasant to look at. Among *'gouramies fish'*, 3 spot gourami, pearl gourami, moon light gourami, snake skin gourami, dwarf gourami, giant gourami, chocolate gourami and kissing gourami are the important species. Bettas and gouramies are the most popular among nest-builders, characterized by possession of accessory respiratory organs, thus making them hardier (Helfman *et al.*, 2009; Madhuri, 2013; MANAGE, 2008; Pandey, 2013b and 2014b).

For species like goldfish which lay adhesive eggs, the matured males and females are selected. Breeding tanks are provided with some artificial net or aquatic plants. Spawning takes place within 6 to 12 hr with adhesive eggs attaching to the plants/net. Parents are removed from the breeding pool or nets/plants with attached eggs are transferred to separate tanks for hatching. The hatching occurs within 72 hr at 27 to 30°C. Nets/plants are removed from the hatching tank subsequently and rearing of larvae is continued in the same tank. In laying non-adhesive eggs like species zebrafish, hatching of eggs takes place in 2 to 3 days after fertilization. The parents are removed from the breeding pool to avoid predation of young larvae. Live-bearers are the group of ornamental fishes which gives birth to young ones. They produce only a few numbers of offspring in comparison to egg-layers. Breeding of most of the live-bearers is relatively easy. Development of baby takes place inside the female and they are released after about 4 weeks. The species of live-bearers include guppy (*P. reticulata*), black mollie (*P. sphenops*), swordtail (*Xiphophorus helleri*) and platyfish (*X. maculatus*). The number of babies produced by a live-bearer is usually less than 100 only, though some larger sword-tails may produce as many as 200 babies. The guppy, platy and sword-tail take 6 to 8 weeks to mature, while the mollies take 12 to

16 weeks. Matured males and females of are selected from the community tanks and stocked inside the breeding traps made of perforated nylon bags placed in separate breeding pools meant for each species. The breeding pools are provided with aquatic plants to facilitate breeding. Fertilization is internal in these live-bearers, where the male inserts its gonopodium to transfer milt into the female and the eggs are fertilized inside the female. Embryos grow within 4 weeks to tiny young ones and become ready for free swimming. Soon after birth, the young ones are collected in order to avoid parental predation (Pandey, 2013b and 2014b).

Foods for Ornamental Fishes

Usually, feed is supplied in aquaculture due to insufficiency of natural food in rearing environment, and for good production. But in case of ornamental fish culture, the feed is supplied not only for their growth but also for keeping them healthy. Feed also helps for the improvement of body colour. The live feeds are generally accepted by the ornamental fishes, and they do not pollute the environment. Some of the live feeds, *viz.*, *Infusoria, Brachionus, Moina* and *Daphnia* can be cultured, and given to both babies and adult fishes. Artificial feed as dried/moist/paste forms can also be given. Dried forms can be of pellets, flakes and freeze dried. The size of feed is to be made perfectly according to fish size. Flakes are more preferred due to their floating nature. Moist feed can also be given, but it hampers the keeping quality. The fishes should be provided feed 4 to 5 times daily for improving the feed utilization efficiency and minimizing the wastage (Pandey, 2013b and 2014b).

Aquarium Plants for Ornamental Fishes

The aquatic plants provide aesthetic beauty to the aquarium and create natural surroundings to the fishes. They reduce the carbon dioxide (CO_2) content of water during illumination and release oxygen (O_2) which is essential for the respiration of fishes. The aquarium plants give shelter to fishes, especially when smaller ones are chased by the larger fishes. Many species of aquarium plants are available with aquarists. Some of them are costlier than even aquarium fishes, indicating their importance in the aquarium tank. Commonly available plants are tall, grass-like plants, *viz.*, *Vallisneria, Sagittaria* and *Acorus*. Bushy plants like *Hygrophila, Ludwigia* and *Ceratopteris* are quite attractive. Amazons sword plant, *Echinodorus* is one of the most sought-after plants by aquarists. *Aponogeton, Cabomba, Ceratophyllum, Hydrilla* and *Myriophyllum* are some of the bunch plants. These plants can tolerate varied water conditions, moderate to strong light and can propagate easily in small and large aquarium tanks. A combination of 4 to 5 of these plants would be ideal for an aquarium (Madhuri, 2013; Pandey, 2013b).

Setting and Maintenance of Aquarium

The aquarium tank could be of different size and shape. It requires general equipment like glass, silicon rubber and squeezing gum for joint, polythene sheet, adhesive tape, sharp knife, glass cutter, scale, etc. Thickness of glass depends on the depth and size of aquarium. Usually, 5 mm thick glass is preferable for 30 cm deep water column. Tanks can be made of glass bonded together with adhesives. Ancillary parts of an aquarium are aerator, filter, decorative toys, aquarium plants, sand and gravels at the bottom, light and heat arrangements, etc. The plantation, sand and gravel give natural habit as well as give descent look to the aquarium. The filter helps in removing the harmful substances, *e.g.*, nitrite, ammonia, etc. produced by the metabolism of fish. Filter also provides suitable, healthy and stable environmental conditions. A good water quality is the basic requirement for providing suitable environment to the fish. Fresh and clear portable water is ideal, but the chlorinated water should be aerated overnight before adding to the aquarium. As over feeding leads to deterioration of water quality siphoning out the left over feed and faecal matter from the bottom of tank or exchange, the water at regular interval is essential for providing congenial living environment. The scum formed on the wall of the tank also requires the regular cleaning. The aquarium with more number of fishes is to be provided with required aerators to keep O_2 level within desirable limit. Because the aquarium fishes are delicate, sensitive and susceptible to many types of diseases, it is essential to have a careful watch daily to control the infections (Helfman *et al.*, 2009; Madhuri, 2013; MANAGE, 2008; Pandey, 2013b).

Common Diseases of Ornamental Fishes

Ornamental fishes are delicate, sensitive and susceptible to various diseases or infections. Diseases/infections may occur due to poor water quality, temperature fluctuation, poor quality feed, etc. General symptoms in diseased fish appear to be the loss of appetite and colour, abnormal swimming behaviour, sluggishness, etc. Common diseases, *e.g.*, columnaris disease, fin rot and tail rot disease, and white spot disease are seen in ornamental fishes. The pop eye disease, velvet disease, slimy skin disease, dropsy, etc. are also noticed in various aquarium fishes. The ectoparasite infection due to fluke attack, fish lice and anchor worm are seen in many cases (Madhuri, 2013; MANAGE, 2008; Pandey, 2013b and 2014b).

Most bacteria that infect the fish are Gram-negative, *e.g.*, *Aeromonas hydrophila*, *A. salmonicida*, *Flavobacterium columnare* (that causes columnaris disease), *Vibrio* and *Pseudomonas* species. Major group of Gram-positive bacteria that cause fish disease is *Streptococcus*. A third group, the acid-fast bacteria includes *Mycobacterium* species (which cause mycobacteriosis). Most bacteria that cause diseases in fish are normal inhabitants of aquatic systems and, ordinarily, they do not cause problem. However, the fish that are stressed by one or more factors (*e.g.*, temperature changes, poor water quality, recent transport or handling), may have compromised (less effective) immune systems, making them more susceptible to bacterial infections. Apart from this, the stress factors which compromise the immune system of fish may actually favour the bacterial growth, further increasing the risk of a disease outbreak (Pandey, 2014b; Yanong, 2014).

Treatment and Health Management of Ornamental Fish

Many disease problems of ornamental fish begin due to external infections. If uncontrolled, the infections may become systemic, resulting in death of the affected fish. Ornamental fish culturists (especially poor ones) use different chemicals or disinfectants and antibiotics to get rid of the disease outbreak. The common '*disinfectants*' used are formalin, malachite green, potassium permanganate and methylene blue. Correct use of KMnO$_4$ can effectively control many bacterial, parasitic and fungal agents before systemic infections become established, often eliminating the need for antibiotic therapy. The fish owner saves money, because the use of expensive antibiotics is decreased, thereby decreasing the incidence of resistant bacterial strains. In ornamental trade, correct use of KMnO$_4$ at the onset of an infection can also speed the movement of fish as they do not need to be held for lengthy (often 10-14 days) antibiotic treatments (Pandey, 2013b). The diseased ornamental fishes must be separated and given a mild *'dip treatment'* with KMnO$_4$ (0.002 per cent), common salt (3 per cent) or CuSO$_4$ (0.05 per cent). Chloromycetin (40 mg/5 L of water) may be added in aquarium tank to prevent some bacterial diseases. Aquarium should be cleaned with disinfectants like KMnO$_4$, malachite green or formalin in case of high disease intensity. Badly infected fishes should be removed to prevent further spreading of disease (Madhuri, 2013; MANAGE, 2008; Pandey, 2013b and 2014b).

Regarding the use of *'antibiotics'* in ornamental fish diseases/infections, the directions are described as follows (Pandey, 2014b; Yanong, 2014):

1. The most common *'antibiotics'* used against bacterial infections are penicillins, oxytetracycline, erythromycin, chloramphenicol, aminoglycosides. Other *'antibacterial agents'*, *e.g.*, sulpha drugs, nitrofurans and quinolones are also used. Most of the ornamental fish farms use chloramphenicol, followed by oxytetracycline and erythromycin. The disinfectants and antibiotics are effective mostly prior to infection, or at the preliminary stages of infection. Mostly, the ornamental fish farmers use prophylactic treatment measures based on eye observations of clinical signs. The disinfectants are used to treat fungal and parasitical diseases; however, the antibiotics are used to treat bacterial diseases and, to some extent, parasitical diseases.

2. *'Food and Drug Administration'* (FDA) has legalized five drugs/chemicals in the US aquaculture that include oxytetracycline, sulphamerazine, ormetoprim, formalin and tricaine methanesulphonate. The FDA has also legalized the following drugs to be used in animals: tricaine methanesulphonate, formalin-F, Paracide-F, Parasite-S, Romet 30, sulphamerazine in fish grade, terramycin-200, Chorulon, hydrogen peroxide, Aquaflor, Aquaflor-CA1, terramycin-343 and oxytetracycline for fish. Chloramphenicol (mostly used in farms) and nitrofurans have been banned worldwide for use in food production because of their serious toxicity. Chloramphenicol may cause fatal aplastic anaemia, while nitrofurans are carcinogens. Besides, antibiotics in the fish farms have the risk of developing bacterial resistant. Malachite green is a respiratory poison. In water, KMnO$_4$ is quickly transformed into non-toxic manganese dioxide, which precipitates out. It is toxic for phytoplankton. Formalin can develop a white precipitate of paraformaldehyde, which is more toxic than pure formalin. Disinfectants can also create pollution in water. Fish farms may use procedure to avoid diseases by regular water quality management, probiotics or other herbal medicines with fewer side-effects. If farms are infected with disease, fish must be treated with approved chemicals/drugs.

3. Ideal solution to bacterial diseases involves working with a fish health specialist to culture the organism and to run sensitivity tests. *'Culture'* refers to growing the infective bacteria on a special type of media (or *'food'*; usually agar or gelatin based), while *'sensitivity'* refers to treating the bacteria with a series of antibiotics to determine which one will work best. Although culture and sensitivity tests generally take 2 or 3 days, they are the best methods for selecting antibiotic that will successfully and economically treat the infection. Fish health specialist should be contacted and given a complete history of the problem during initial stages of disease. The specialist will provide instructions on submitting samples to a diagnostic laboratory. The affected fish should not be treated with any antibiotic until after a sample has been analyzed. The sample should include at least 3 to 5 fishes, showing typical signs of disease. The fishes that are submitted after they have been given antibiotics often provide poor culture results. While waiting for culture results, the fish health specialist may suggest a broad-spectrum antibiotic that can be used until culture and sensitivity tests have been completed. Legalities must also be considered when selecting antibiotics. Producers of food fish have fewer options than producers working with ornamental fish because the FDA has elected to use regulatory discretion to keep products intended for the treatment of aquarium fish available. The fish health specialist will be able to provide information on legal constraints for specific antibiotics, information for correct dosages, proper methods of administration and concerns about effluent from ponds with treated fish. Antibiotics, in and of themselves, do not cure a fish. They merely control the population growth of bacteria in a fish long enough for its immune system to eliminate them. Before antibiotics are considered, sources of stress like poor water quality (including drastic temperature change), nutrition, genetics and handling or transport must be removed or reduced. The affected fish should also be examined for parasites. Any of these factors may be the primary cause of disease, as bacterial infections are often secondary responses to such management problems. Contacting a fish health specialist early during disease outbreak will help identify contributing stressors and the rate of bacterial infection, so the total loss of fish will be reduced.

4. Injection is the most direct and effective method for getting antibiotics into the blood stream. Unfortunately, this process is very labour intensive and impractical for fish grown on a commercial scale. However, for small numbers of fish, or for important or expensive fish, injection may be the best method. In aquaculture production, the most cost effective and commonly used method to deliver antibiotics is *'orally by mixing them into food'*. The proper dose of antibiotic is mixed into the feed during production, or else it is added after production, using fish oil or canola oil as a binding agent. The mixture is then fed to fish for the prescribed number of days. Oral administration of antibiotics requires that most of the fish are still eating, so every attempt should be made to catch bacterial diseases early, before

the majority of fish stops eating. Only fishes that are eating will be treated. Although *'bath treatments'* are a popular method of administering antibiotics, much more drug is required to achieve the desired result as compared to oral treatments or injections. In many cases, even a large amount of antibiotic in the water does not guarantee that enough of it will get into the fish. At the same time, excessive amounts of antibiotic in the water can increase the likelihood of water-borne bacteria, developing the resistance to that drug. Further, to avoid poor water quality and any potential toxicity, between 70 per cent and 100 per cent of the water should be changed at the end of each daily treatment and also prior to re-dosing. Finally, bath treatments are not recommended in re-circulating systems or in any aquarium system where the treated water will contact the biological filter, because the antibiotics may kill or inhibit the nitrifying bacteria in the biological filters. In conclusion, bath treatments should be considered only when the majority of fishes are not eating, or when treating primarily external bacterial infections and fishes should be switched to oral medications as soon as they resume eating. Thus, oral/food treatments are more effective, more efficient, less detrimental to water quality and preferable to bath treatments.

5. *'Erythromycin'* is most effective against Gram-positive bacteria, *e.g.,* *Streptococcus* species. It is not very effective in a bath treatment, and it should only be administered by injection or in feed. It is not FDA-approved for use with food fish. Oral (food) dose of erythromycin is 1.5 g per lb food per day for 10 days.

6. *'Penicillins, including penicillin, amoxicillin and ampicillin'* are most effective against Gram-positive bacteria like *Streptococcus* species. Hence, these are not the first choice for most bacterial infections in fish (as erythromycin is given for the same reason). None of the penicillins are FDA-approved for use in food fish. The respective oral (food) doses of amoxicillin and ampicillin are 1.2 to 3.6 g and 150 mg per lb food per day for 10 days.

7. *'Oxytetracycline and related antibiotics'* are broad-spectrum antibiotics (effective against a wide variety of bacteria) which act well when mixed with food; however, their bath treatments may not be as effective for all species. Calcium and magnesium bind to oxytetracycline and tetracycline, rendering them inactive. This means that with increasing water hardness (*i.e.,* increases in calcium and magnesium levels), it is necessary to increase the doses of these drugs in bath treatments. *'Tetracyclines'* are ineffective when used as a bath treatment for saltwater fish. They are light sensitive and turn brown when decomposing. This contributes to poor water quality and may be harmful to the fish. Water should be changed immediately after the bath treatment period is concluded. Due to years of misuse, several bacteria in many different facilities are now resistant to tetracyclines. Oxytetracycline still works adequately against most cases of *F. columnare* (columnaris disease). *'Terramycin'*, a brand of oxytetracycline, is FDA-approved for use in the production of salmonids, channel catfish and lobsters. The oral (food) dose of oxytetracycline (terramycin) is 1.12 g per lb

food per day for 10 days; while its bath dose (water change recommended prior to each dosing) is 750 to 3,780 mg per 10 gallons for 6 to 12 hr (repeat daily for 10 days; dose will depend on the hardness of water).

8. *'Aminoglycosides, including gentamicin, neomycin, kanamycin and amikacin'* are very effective against Gram-negative bacteria when given by injection. However, they cause kidney damage in fish when given by this route. These antibiotics are not effective when used in oral or bath treatments. Exceptionally, kanamycin and neomycin may be effective against external infections if used in bath treatments. Besides, kanamycin is also effective when mixed with feed to treat gastrointestinal bacterial infections. None of the aminoglycosides are the FDA-approved drugs for use with food fish. The respective oral (food) doses of gentamicin and kanamycin are 45 and 300 mg per lb food per day for 10 days. The bath dose (water change recommended prior to each dosing) of kanamycin is 189 to 378 mg per gallon, every 3 days for three treatments. The oral (food) dose of neomycin is 1.5 g per lb food per day for 10 days; while its bath dose (water change recommended prior to each dosing) is 2.5 g per 10 gallons, every 3 days for three treatments.

9. *'Florfenicol'* is a broad-spectrum antibiotic that is FDA-approved in fish under the trade name *'Aquaflor'*. This is sold as a *'veterinary feed directive drug'*, that requires a veterinary prescription for use and that cannot be used outside of its label claim. It is approved against specific diseases in catfish, freshwater-reared salmonids and freshwater-reared finfish (including freshwater-reared warm water finfish). For a 1 per cent feeding rate, florfenicol (Aquaflor) is given @ 680 mg per lb, orally for 7 to 10 days.

10. *'Sulpha drugs, including Romet'* are also considered to be broad-spectrum antibacterial agents. There are many types of sulpha drugs. The more common sulpha drugs are not as effective as they once were, due to their misuse or overuse, resulting in the creation of many bacteria that are now resistant to them. On the other hand, the potentiated sulpha drugs, including Romet are still very effective. Romet acts well when mixed with feed, but it does not act well as a bath treatment. It is FDA-approved for use with channel catfish and salmonids. The oral (food) dose of Romet (sulphadimethoxine/ormetoprim) is 0.148 lb per 20 lb food per day for 5 days.

11. *'Nitrofurans, including nitrofurantoin, nitrofurazone, furanace and furazolidone'* are commonly used in ornamental fish trade, but the FDA strictly forbids their use with food fish. Nitrofurazone is specifically categorized as *'of high regulatory concern'* by the FDA, and should not be present at any facility where food species are raised. Although nitrofurans are commonly used in bath treatments, they are probably most effective against superficial infections. Within the group, furanace appears to be the most effective for use in bath treatments. To achieve the most effective result when using a nitrofuran in a bath treatment, either some sort of cover or a darkened treatment location should be considered, as the nitrofurans can be

inactivated by light. Nitrofurans as oral treatments may not be as effective as previously believed. The oral (food) dose of nitrofurazone is 1.12 g per lb food per day for 10 days; while its bath dose (water change recommended prior to each dosing) is 189 to 756 mg per 10 gallons for 1 hr (repeat daily for 10 days) or 378 mg per 10 gallons for 6 to 12 hr (repeat daily for 10 days).

12. *'Quinolones, including nalidixic acid and oxolinic acid'* are also broad-spectrum antibacterial agents, like tetracyclines. They act against a wide variety of bacteria. They act best at acidic pH of 6.9 or less, and are inhibited by hard water. Although they appear to work well in both bath and oral treatments, some fishes may sink to the bottom and appear lethargic after bath treatments. They damage the nervous system of other animals, and none are approved by the FDA for use with fish. Quinolones are closely related to a category of antibacterials, called *'fluoroquinolones'*, which are categorized as *'of high regulatory concern'* by the FDA. Use of fluoroquinolones or quinolones to treat any food animal is illegal and completely irresponsible. The oral (food) dose of nalidixic acid is 300 mg per lb food per day for 7 to 10 days; while its bath dose (water change recommended prior to each dosing) is 500 mg per 10 gallons (repeat as needed). However, the oral (food) dose of oxolinic acid is 150 mg per lb food per day for 10 days; while its bath dose (water change recommended prior to each dosing) is 38 mg per 10 gallons for 24 hr (repeat as needed) or 95 mg per gallon for 15 minutes (repeat twice daily for 3 days).

Chapter 4

Medicinal Plants Acting Against some Fish Diseases

Medicinal Plants for the Treatment of Fish Diseases

Role of Medicinal Plants in Fish Diseases

Treatment with medicinal plants having antibacterial activity is a potentially beneficial alternative in the aquaculture. These herbs mitigate many of the side effects, which are associated with synthetic antimicrobials. Besides, the plant-derived phytomedicines provide a cheaper source for treatment and greater accuracy than the chemotherapeutic agents. Plants have been used as traditional medicine since time immemorial to control bacterial, viral and fungal diseases. Recently, the research has been initiated to evaluate the feasibility of herbal drugs in fish diseases (Madhuri *et al.*, 2012a and 2012i; Pandey, 2013c).

Although herbal remedies have been with us for human therapy for millennia, there has been relatively little research on the medicinal plants to be used against fish diseases. Medicinal plants (herbal drugs) can be used not only as remedies but even more so, as growth promoters, stress resistance boosters and preventatives of infections. Hence, herbal drugs in disease management are gaining success, because they are cost effective, eco-friendly and have minimal side effects. A large portion of the world population, especially in developing countries depends on the traditional system of medicine for a variety of diseases. Several hundred genera are used medicinally and plants are vital sources for potent and powerful drugs. Plants are rich in a wide variety of secondary metabolites of phytochemical constituents, *e.g.*, tannins, alkaloids and flavonoids, which act against different diseases (Pandey and Madhuri, 2010a; Pandey *et al.*, 2012a; Ravikumar *et al.*, 2010).

Diseases caused by *A. hydrophila* bacterium are some of the most widespread in freshwater fish culture. Septicaemia caused by motile aeromonads is a ubiquitous problem that affects fish found in warm, cool and cold freshwater around the world. *A. hydrophila* has been associated with diseases in fishes like carp, eels, milkfish, channel catfish, tilapia and ayu. This microorganism can also be an opportunist in stress-related diseases in salmonids. Antibiotics are frequently used to control disease caused by this bacterium, but there is an increasing risk of developing antibiotic resistant (Yin *et al.*, 2008). *A hydrophila* is also responsible for skin infections, septicemia and gastroenteritis in human, besides the fish (Castro *et al.*, 2008). Continuous use of antimicrobial agents in aquaculture has resulted in more resistant bacterial strains. Since ancient times, medicinal plants have been used for the treatment of common infectious diseases; and the treatments with plants having antibacterial activity are a potentially beneficial alternative in aquaculture. In addition, the phytomedicines provide a cheaper source for treatment and greater accuracy than chemotherapeutic agents in fish (Abdul Kader Mydeen and Haniffa, 2011; Turker *et al.*, 2009). *A. hydrophila*, the most common bacterial pathogen in freshwater fish, has been recognized to be the aetiological agent of several distinct pathological conditions, including tail rot, motile Aeromonas septicemia (MAS) and epizootic ulcerative syndrome (EUS) as a primary pathogen. EUS is a globally distributed disease and has become an epidemic affecting a wide variety of wild and cultured fish species in South-east Asia, including India. Ability of herbs to inhibit activity of bacteria having potential interest as fish pathogens has been documented. Heavy antibiotics in aquaculture should be reduced and replaced with alternative processes in fish diseases to avoid the emergence of antibiotic resistance in pathogenic and environmental bacteria (Abdul Kader Mydeen and Haniffa, 2011). The phytochemicals, *e.g.*, tannins, alkaloids and flavonoids present in the medicinal plants have been found *in vitro* to have antimicrobial properties. Many of the spices and herbs have been valued for their antimicrobial effects and medicinal powers in addition to their flavour and fragrance qualities. Research has been initiated to evaluate the feasibility of using herbal medicines in fish disease management. Moreover, bacterial infections are considered the major cause of mortality in aquaculture; while the medicinal plants and their products play significant role in fish culture (Ravikumar *et al.*, 2010).

Chinese herbs have been used as medicine to treat different fish diseases in China for many years. Herbs have also been used in other countries for control of shrimp and fish diseases, and successful results have been reported in Mexico, India, Thailand and Japan (Pandey, 2013c; Pandey *et al.*, 2012a; Yin *et al.*, 2008). Traditional herbal medicines seem to be the potential immunostimulator. Thus, the use of medicinal plants is an alternative to antibiotics in fish health management. Herbs are not only safe for consumers, but they also have significant role in aquaculture. Many studies have proved that herbal additive enhanced the growth of fish and protected from disease. Non-specific immune system of fish is considered to be the first line of defense against invading pathogens (Ahilan *et al.*, 2010). Many plant-compounds have been found to have non-specific immunostimulating effects in humans and animals, of which more than a dozen have been evaluated in fish and shrimp. Several plant products seemed to be potent antiviral agents against fish and shrimp viruses (Pandey *et al.,* 2012a). *Hence,* the medicinal plants may be used as potential and promising source of pharmaceutical agents against fish pathogens in organic aquaculture (Abdul Kader Mydeen and Haniffa, 2011; Pandey *et al.,* 2012a; Ravikumar *et al.,* 2010).

Certain Medicinal Plants for Treatment of Fish Diseases

To preserve and protect the environment as well as human health as a best alternative, different parts of *Azadirachta indica* (Neem) tree have been studied. Neem leaves containing nimbin, azadirachtin and meliantroil have been reported to possess a variety of properties, including insecticidal and antiviral from ancient times. Indian almond (*Terminalia catappa*) and garlic (*Allium sativum*) have been said as an alternative to chemicals to treat fish ectoparasites, *Trichodina* species infections in tilapia (*O. niloticus*) fingerlings. Both Indian almond and garlic had low acute toxicity to tilapia fingerlings, treating the trichodiniasis caused by *Trichodina*. Further, the Indian almond, commonly used as herb in Taiwan, prevents the fish diseases. It is claimed to be a wound healing substance for *'Siamese fighting fish'* hurt after matches in Thailand as well. The immunostimulant effects of the dietary intake of 3 plants (*viz.,* *Viscum album, Urtica dioica* and *Zingiber officinale*)-extracts on rainbow trout (*Oncorhynchus mykiss*) have also been observed (Chitmanat *et al.,* 2005). Christybapita *et al.* (2007) noticed the immunostimulatory effect of aqueous extract (AqE) of *Eclipta alba* (Bhangra) leaf (oral administration as feed supplement) in tilapia fish, *Oreochromis mossambicus*. It was noted that *E. alba* extract enhances non-specific immune responses and disease resistance of *O. mossambicus* against *A. hydrophila* infection. According to Winkaler *et al.* (2007), *A. indica* extract can be used successfully in aquaculture to control fish predators.

Castro *et al.* (2008) screened the methanolic extracts of 46 Brazilian plants and found only 31 to have the antibacterial activities against fish pathogenic bacteria, *viz., Streptococcus agalactiae, F. columnare* and *A. hydrophila*. Yin *et al.* (2008) added the extracts of two Chinese herbs (*viz., Lonicera japonica* and *Ganoderma lucidum*) in diets of tilapia fish (*O. niloticus*), and found that these herbs act as immunostimulants and appear to improve the immune status and disease resistance. Both herbs when used alone or in combination increased the survival of fish after challenge with *A. hydrophila*. On the basis of several studies, Yin *et al.* (2008) reported that oral administration of ginger (*Z. officinale*) extract increases the phagocytic capability of cells in rainbow trout; while the extracts of four Chinese herbs (*Rheum officinale, Andrographis paniculata, Isatis indigotica* and *L. japonica*) increased the phagocytosis of white blood cells (WBCs) of carp. Turker *et al.* (2009) reported that the alcoholic and aqueous extracts of *Nuphar lutea, Nymphaea alba, Stachys annua, Genista lydia, Vinca minor, Fragaria vesca, Filipendula ulmaria* and *Helichrysum plicatum* herbs of Bolu (Turkey) have antibacterial activity against *A. hydrophila, Yersinia ruckeri, Lactococcus garvieae, S. agalactiae* and *Enterococcus faecalis* bacteria isolated from fish. This observation provided the aquaculturists with a promising management tool for control or treatment of fish diseases. Nya and Austin (2009) observed the control of *A.*

hydrophila infection after feeding with *A. sativum* (0.5 and 1 g/100 g of feed for 14 days) to rainbow trout, *O. mykiss*.

Ahilan *et al.* (2010) observed that the addition of *Phyllanthus niruri* and *Aloe vera* (Aloe) as herbal additives can positively enhance the growth performance of goldfish (*Carassius auratus*) as well as its resistance to *A. hydrophila* infection. Further, the herbal additives in diets often provided cooperative action to various physiological functions. The synergistic effect of herbs was reported in other fishes, including Japanese flounder and *Clarias gariepinus*. The growth increase in *Labeo rohita* fish fed with herbal supplemented diet was due to improved food utilization and high protein synthesis. The benefit of herbal growth promoters as an additive in the carp feed was also found. Furthermore, the medicinal plants, *viz.*, ginger, nettle and mistletoe as an adjuvant therapy in rainbow trout through feed enhanced the phagocytosis, and cellular and humoral defense mechanisms against pathogens. Traditional Chinese medicines in yellow croaker elevated the non-specific defense mechanism and increased disease resistance of fish against bacterial pathogens. The disease resistant of *Catla catla* fish was produced through immersion treatment of three herbs, *viz.*, *A. sativum*, *A. indica* and *Curcuma longa* (Haldi, turmeric) in spawn. *A. vera* has been found as a disease suppressing agent and showed antibacterial effect in juvenile rock fish. Harikrishnan *et al.* (2010) reported that mixed herbal extracts diets restored the altered haematological parameters and triggered innate immune system of goldfish (*C. auratus*) against *A. hydrophila* infection.

Chloroform extract of *Datura metel* plant has wide range of antimicrobial activity against many fish pathogens. This plant can be used as a putative antimicrobial drug in aquaculture maintenance. Chloroform extract of *D. metel* can be used as a potential antimicrobial agent to overcome the problem of mass mortality of ornamental fish in aquarium so as to enhance the market revenue throughout the world. Antimicrobial activity of five Chinese herb extracts against 13 bacterial and two viral fish pathogens was also seen (Ravikumar *et al.*, 2010). Sharma *et al.* (2010) noted the stimulatory effect of dietary doses of *W. somnifera* root on immunity and disease resistance against *A. hydrophila* infection in *L. rohita* fingerlings. Abdul Kader Mydeen and Haniffa (2011) cited that *A. indica* leaf AqE could effectively control *A. hydrophila* infection in common carp (a freshwater fish usually bred in ponds), *C. carpio*. Further, *Enterobacter* species and *Escherichia coli* bacteria, isolated from marine fish (*Amphiprion sebae*) showed 15 mm zone of inhibition against neem extract. The antimicrobial activity of AqE of three medicinal plants, *viz.*, *A. indica* (leaf), *Solanum torvum* (Sundakai fruit coat) and *C. longa* (rhizome) against the *in vitro* growth of *A. hydrophila*, isolated from infected freshwater fish, *Channa striatus*, was also noticed by these workers. Nargis *et al.* (2011) observed immunostimulant effects of the dietary intake of *A. sativum* and *Vitex negundo* extracts on fingerlings of *L. rohita* fish. Ravikumar *et al.* (2011) studied that among 15 coastal medicinal plants/parts of plants, *A. indica*, *Cinnamomum verum* and *Eupatorium odoratum* exhibited excellent antibacterial activity against 10 bacterial pathogens from diseased ornamental fishes. Some herbal extracts are very effective against gills and skin flukes like *Benedenia seriolae* (Pandey *et al.*, 2012a).

Herbal Drugs Against Fish Ectoparasites

Elimination of Fish Ectoparasites by Herbal Drugs

Treatment for ectoparasitic diseases in freshwater fish with formalin seems at present to be ineffective. For this reason, it is evidently a useless cost. In addition, formalin possibly leaves toxic residues in fish flesh and in the environment which are eventually harmful to consumers. The alternative way to solve this problem is to use traditional medicinal plants instead (Chitmanat *et al.*, 2005; Pandey, 2013c).

Unfortunately, the parasitic outbreak acts as an important limiting factor for aquaculture business. Every tilapia fish (*O. niloticus*) reared by the farmers in Chiang Mai (Thailand) has a *Trichodina* parasite infection. This problem was significantly related to the high organic matter in water (Pandey, 2013c; Pandey *et al.*, 2012a). Chitmanat *et al.* (2005) reported that heavy infection of *Trichodina* species in small fish caused gigantic financial losses. The infected fish are lethargic, generate excessive mucus and become off-feed eventually, resulting into considerable deaths. There is now a fast growing interest in screening antiparasitic substances from plants to replace chemical and antibiotic alternatives.

Garlic and Indian almond have been reported (Chitmanat *et al.*, 2005) to treat tilapia fish ectoparasites, *viz.*, *Trichodina* species. Tilapia is one of the many economical freshwater fish that are cultured worldwide. In 2003, the increased export of tilapia brought in almost 146 millions Thai baht. Trichodiniasis used to be effectively controlled with formalin, but it now appears insufficient to control this parasitic infection. This leads to increased needless cost and possibly causes undesired toxic residues in fish flesh and in the environment (Pandey, 2013c).

Heavy antibiotics in aquaculture should be reduced and replaced with alternative processes in fish diseases to avoid the emergence of antibiotic resistance in pathogenic and environmental bacteria (Abdul Kader Mydeen and Haniffa, 2011). Food safety is a great concern worldwide, so the producers need to generate clean food products. There is a fast growing interest in screening antiparasitic substances from plants to replace chemical and antibiotic alternatives. Many such plants have been reported to act against ectoparasites. For example, a wide range of microorganisms including bacteria, fungi, protozoa, and viruses have been shown to be sensitive to crushed garlic preparations (Chitmanat *et al.*, 2005).

Some Herbal Drugs Acting Against Fish Ectoparasites

Punicalin from the leaves of Indian almond inhibited HIV replication in infected H9 lymphocytes with little cytotoxicity. Using the crude extract from either garlic or Indian almond is one of the new challenging methods for trichodiniasis treatment. Allicin, one of the active ingredients in garlic, exhibits antiparasitic activity, including some major intestinal protozoan parasites like *Entomoeba histolytica* and *Giardia lamblia*. The raw and squeezed garlic at 200 mg per L has potential to treat trichodiniasis in eel fish. Indian almond also prevents various fish diseases. It is commonly used as an herbal drug in Taiwan. It has been locally claimed to be a wound healing substance for Siamese fighting fish hurt after matches in Thailand as well. The crude ethanol extract from the leaves of Indian almond was found to be active against *Staphylococcus aureus* bacterium with a minimum inhibitory concentration of 512 g per ml. In addition, the crude extract of Indian almond has *in vitro* antifungal properties against *Pythium ultimum, Rhizoctonia solani, Sclerotium rolfsii* and *Aspergillus fumigatus* (Pandey, 2013c).

Chitmanat *et al*. (2005) observed that the crude extracts of either garlic or Indian almond at 800 mg per L significantly (P<0.05) eliminated the *Trichodina* species infections in tilapia (average weight 3.62±0.06 g each). Either garlic or Indian almond at 800 ppm was able to remove all *Trichodina* species from tilapia after 2-day treatment. All treatments were significantly different from control groups. Both garlic and Indian almond are more economical than formalin and, for this reason the cost of treatments would be reduced. However, *Trichodina* species became re-apparent after two weeks. As both Indian almond and garlic contain high organic matter, they are possibly able to cause water deterioration. It is suggested to exchange the water a day after treatment. Based on this study, either garlic or Indian almond could be developed for safer ectoparasitic treatment. As it is extremely desirable to reduce the use of hazardous therapeutics for trichodiniasis control, there is a great potential of using garlic and Indian almond for this parasitic treatment. Both garlic and Indian almond had low acute toxicity to tilapia fingerlings at the working concentration to treat trichodiniasis.

The infection pressure from trichodiniasis in farms with a relatively high load of organic matter (>15-20 mg/L) may be relieved by reducing the content of organic dry matter in the processed water. Some other medicinal plants have been used as antibiotic and chemical alternatives. The rainbow trout fish fed with diets, containing AqE of mistletoe, nettle and ginger, exhibited the significant non-specific immune responses. Since all medicinal plants are able to stimulate only non-specific immune responses, vaccines might be a better way to prevent the deadly diseases. These plants could be used as vaccine adjuvants though in order to create more effective vaccines (Pandey, 2013c).

Medicinal Plants Against Fish Microbial Pathogens

Bacterial Infections in Fish

Bacterial infections are the major cause of mortality in aquaculture (Madhuri *et al., 2012a*). Among common fish pathogens, *A. hydrophila* and *Y. ruckeri* as Gram-negative, and *S. agalactiae, L. garvieae* and *E. faecalis* as Gram-positive bacteria cause infectious diseases (Abdul Kader Mydeen and Haniffa, 2011; Turker *et al.*, 2009). *S. agalactiae, L. garvieae* and *E. faecalis* are closely related groups of bacteria that can cause diseases like streptococcosis, lactococcosis, haemorrhagic septicemia and ulcers in fins. *F. columnare* is pathogenic only to freshwater fish species and shows low environmental fitness, when compared with other aquatic bacteria. Even though, this agent is highly virulent to young fish (fry and fingerling), causing skin lesions and high mortality, usually associated with poor environmental conditions. Fish are susceptible to several bacterial infections, mainly when reared in high density conditions. Disease outbreaks elevate the mortality rate and decrease productivity efficiency, causing high economic loss of fish farmers. Due to use of a wide variety of antibiotics, aquaculture has been implicated as potential environment to development and selection of resistant bacteria and a source of these pathogens to other animals and humans (Madhuri *et al.*, 2012a).

Some bacterial fish pathogens are also associated to diseases in humans, making the aquaculture products a potential risk to the customers (zoonotic or food borne diseases). *S. agalactiae* is a dangerous pathogen to freshwater and marine fish. The infection is characterized by brain invasion, nervous signs and septicemia. These bacteria can infect humans, causing mainly pneumonia and meningitis in newborns. Enteric red mouth disease mostly restricted to salmonids is caused by *Y. ruckeri* and reddening of mouth and throat is the most common symptom. *A. hydrophila* is responsible for cases of skin infections, septicemia and gastroenteritis in fish and human. *A. hydrophila*, the most common bacterial pathogen in freshwater fish, has been recognized to be the aetiological agent of several distinct pathological conditions including tail/fin rot, MAS or haemorrhagic septicemia and as a primary pathogen. EUS is a globally distributed disease and has become an epidemic affecting a wide variety of wild and cultured fish species, especially in Southeast Asia including Pakistan and India (Abdul Kader Mydeen and Haniffa, 2011; Madhuri *et al.*, 2012a; Turker *et al.*, 2009)

Treatment of Bacterial Diseases by Herbal Drugs

Continuous use of antimicrobial agents in aquaculture has resulted in more resistant bacterial strains in the aquatic environment. Large-scale settings of aquatic animal husbandry have resulted in an increased antibiotic resistance in bacteria potentially pathogenic to fish and related environment. Heavy antibiotics used in aquaculture needs to be reduced and replaced with alternative process for treating fish diseases to avoid the emergence of antibiotic resistance in pathogenic and environmental bacteria. Occurrence of antibiotic resistant strains of bacteria has been described in aquaculture systems. Regarding the problem of microbial resistance, there is an urgent need to establish the rules to the rational use of antibiotics and the discovery of new drugs and alternative therapies to control bacterial diseases (Castro *et al.*, 2008; Madhuri *et al.*, 2012a).

Treatment of bacterial diseases with different herbs has been safely used in organic agriculture, veterinary and human medicine, and treatments with medicinal plants having antibacterial activity are a potentially beneficial alternative in aquaculture. Since ancient times, medicinal plants have been used for the treatment of common infectious diseases. Medicinal plants as the alternative agents are effective to treat the infectious diseases and mitigate many of side effects that are associated with synthetic antimicrobials. Additionally, the plant-derived phytomedicines provide a cheaper source for treatment and greater accuracy than chemotherapeutic agents in this field. The use of alcoholic extracts of herbs may be suggested for the natural administration of antibiotics effective in fish disease control. The ability of some herbs and seaweeds to inhibit bacteria having potential interest as fish pathogens has been documented. Some of the local herbs and desert plants were reported to inhibit the pathogenic bacteria in aquaculture and referred to limited number of plant species (Madhuri *et al.*, 2012a; Turker *et al.*, 2009).

Herbal drugs are rich in a wide variety of secondary metabolites like tannins, alkaloids and flavonoids, which have antimicrobial properties. Many of the spices and herbs used today have been valued for their antimicrobial effects and medicinal powers in addition to their flavour and fragrance qualities. In India, 500 medicinal plant species are used to pathogenic bacteria. Plants have been used as traditional medicine since time immemorial to control bacterial, viral and fungal diseases. In the recent years, herbs and herbal products plays significant role in fish culture. The usage of heavy antibiotic in aquaculture field needs to be reduced and replaced with alternative process for treating fish diseases (Ravikumar *et al.*, 2010). Thus, due to increasing the resistance of microorganisms to antibiotics and the cost of modern allopathic medicines, the scientists are now looking for medicinal plants, because most of them are safe, costless and effective against a wide range of antibiotic resistant microorganisms like bacteria and fungi (Madhuri *et al.*, 2012a).

Certain Antimicrobial Medicinal Plants Against Fish Pathogens

The antimicrobial activity of five Chinese herb extracts against 13 bacterial and two viral fish pathogens has been reported long back (Madhuri *et al.*, 2012a). The AqE of *A. indica* (Neem) leaf was tested against *A. hydrophila* infection in common carp, *Cyprinus carpio* and the results showed that this plant could effectively control the *A. hydrophila* infection (Harikrishnan *et al.*, 2003). In a study of 46 methanolic extracts of Brazilian plants, 31 extracts were found to exhibit the antibacterial activity against fish pathogenic bacteria, *viz.*, *S. agalactiae*, *F. columnare* and *A. hydrophila*. The *F. columnare* microorganism was the most susceptible to many tested extracts. In contrast to its high virulence to young fish, this bacterium is sensible to the main disinfectants used in fish farms, like $KMnO_4$, hydrogen peroxide, chloramines and salt. Despite of their common use, these compounds may be dangerous to fry and aquatic environment. The plant extracts can be applied as an alternative to prevent and control the outbreaks of columnaris disease, mainly in hatchery. Since these substances are natural, their hazardous potential is lower when compared with other products. The results showed that the analyzed plants presented a high potential as alternative therapy of bacterial fish diseases (Castro *et al.*, 2008).

The alcoholic and aqueous extracts from 22 species of herbs from Bolu (Turkey) were screened for antibacterial activity against *A. hydrophila*, *Y. ruckeri*, *L. garvieae*, *S. agalactiae* and *E. faecalis*. The extracts with various solvent of *Nuphar lutea*, *Nymphaea alba*, *Stachys annua*, *Genista lydia*, *Vinca minor*, *Fragaria vesca*, *Filipendula ulmaria*, *Helichrysum plicatum* showed the highest inhibitory activity. The alcoholic extract of *V. minor*, and the alcoholic and aqueous extract of *N. lutea* displayed a broad antibacterial spectrum against the target organisms (Turker *et al.*, 2009). *Enterobacter* species and *Escherichia coli* isolated from marine fish, *Amphiprion sebae* caused 15 mm zone of inhibition against Neem leaves extract (Dhayanithi *et al.*, 2010). *In vitro* antibacterial and antifungal activity of the chloroform extracts of the 17 different coastal medicinal plants were screened against ornamental fish pathogens, *e.g.*, different Gram-positive and Gram-negative bacteria, and fungi. Of the selected plants, *Datura metel* (Thorn-apple) showed wide range of antimicrobial activity against many of the fish pathogens. *D. metel* showed maximum sensitivity (19 mm diameter) against

Pencillium restrictum fungal species and *Lantana camara* showed maximum sensitivity (13 mm diameter) against *Vibrio* bacteria species. The strong antimicrobial activity of *D. metel* might be due to the presence of various phytoconstituents like alkaloids, flavonoids, phenols, tannins, saponins and sterols. Therefore, *D. metel* plant can be used as a putative antimicrobial drug in the aquaculture maintenance (Ravikumar *et al.*, 2010).

The antimicrobial potency of AqE of *A. indica* (leaf), *Solanum torvum* (Sundakai fruit coat) and *C. longa* (rhizome) against *in vitro* growth of pathogenic bacterium, *A. hydrophila* isolated from infected freshwater fish, *Channa striatus* was assessed. The strongest antibacterial activities among all plants were obtained by the AqE of *A. indica* with inhibition zone of 18 mm against *A. hydrophila*. The *S. torvum* demonstrated moderate (11 mm) and *C. longa* marked weak (8 mm) inhibiting activity against *A. hydrophila*. It was proved that among the three herbs, extract of *A. indica* is very effective against *A. hydrophila* (Abdul Kader Mydeen and Haniffa, 2011). Fifteen coastal medicinal plants/parts were investigated to evaluate the antibacterial activity against bacterial fish pathogens. Among the plants, *A. indica*, *Cinnamomum verum* and *Eupatorium odoratum* exhibited excellent antibacterial activity against 10 bacterial pathogens from diseased ornamental fishes under captivity. *E. odoratum* inhibited the growth of *Vibrio* species (14 mm). Hence, these coastal medicinal plants of Kanyakumari coast possess potential antimicrobial compounds to cure the ornamental fish bacterial diseases (Ravikumar *et al.*, 2011).

Immunostimulatory Action of Herbal Drugs in Fishes

Role of Herbal Drugs as Immunostimulants in Fish

Many studies have shown that herbal drugs or additives enhance the growth of fishes and protect them from diseases. Non-specific immune system of fish is the first line of defense against invading pathogens (Ahilan *et al.*, 2010). Inclusion of herbal drugs or additives in diets often provides cooperative action to various physiological functions. Beneficial role of vitamins C and E has been noted in fish nutrition, reproduction, growth and related indices. In addition, vitamins C and E are credited with modulating the stress response in fish. Biological role played by vitamins C and E is very vital for the sustained growth and health of many living organisms as well as fish. Dietary vitamins have antibody enhancement effects in fish. The synergistic effect of herbal drugs has been reported in many fishes, including Japanese flounder and *C. gariepinus*. The beneficial utility of herbal growth promoters as an additive in the carp feed has been observed. There is a significant difference between different herbal additives on the effect of growth rate in goldfish. Neutrophils and phagocytes, lysozyme and complement are some important indices of nonspecific immunity in fish (Pandey *et al.*, 2012c).

Herbal drugs (medicinal plants) can act as immunostimulants, conferring early activation to the non-specific defense mechanisms of fish and elevating the specific immune response (Pandey, 2013c; Pandey *et al.*, 2012a; Yin *et al.*, 2008). Herbal drugs have been used as medicine and an immune booster for humans for thousands of years. The herbs contain many immunologically active components such as polysaccharides, organic acids, alkaloids, glycosides and volatile oils, which can enhance immune functions. Recently, there has been increased interest in the immune stimulating function of some herbs in aquaculture. The non-specific immune functions such as bacteriolytic activity and leukocyte function of fish have been improved by some mixtures of Chinese herbs in shrimp (*Penaeus chinensis*) and tilapia (Chansue *et al.*, 2000). The non-specific defense mechanisms of fishes include neutrophil activation, production of peroxidase and oxidative radicals, together with initiation of other inflammatory factors (Ainsworth *et al.*, 1991). Moreover, bacterial infections are the major cause of mortality in aquaculture and the medicinal plants and their products play significant role in fish culture (Madhuri *et al.*, 2012b; Pandey, 2013c; Pandey *et al.*, 2012a; Ravikumar *et al.*, 2010).

Fish have a variety of defense to prevent the diseases and parasites. The non-specific defenses include the skin and scales, as well as the mucus layer secreted by the epidermis which traps and inhibits the growth of microorganisms. If pathogens breach these defenses, fish can develop an inflammatory response which increases blood flow to the infected region and delivers WBCs that attempt to destroy pathogens. The specific defenses response to particular pathogen recognized by the fish's body, *i.e.*, to an immune response. During recent years, vaccines have become widely used in aquaculture and also with ornamental fish, *e.g.*, the furunculosis vaccines in farmed salmon and koi herpes virus in koi. Some species of fish use the '*cleaner fish*' to remove external parasites. The best known of these are the '*blue streak*' of the genus *Labroids* found on coral reefs in the Indian and Pacific oceans. These small fishes maintain so-called the '*cleaning stations*', where other fishes congregate and perform specific movements to attract the attention of the cleaners. The cleaning behaviours have been observed in a number of fish groups, including an interesting case between two cichlids of the same genus, *Etroplus maculates* (cleaner) and the much larger fish, *Etroplus suratensis* (Madhuri *et al.*, 2012a; Pandey, 2013c).

Some Medicinal Plants with Immunostimulatory Action in Fish

Eclipta alba leaf (oral administration as feed supplement) was studied for its immunostimulatory activity in tilapia fish, *Oreochromis mossambicus*. Fishes were fed for 1, 2 or 3 weeks with diets containing AqE of *E. alba* leaves at 0, 0.01, 0.1 or 1 per cent levels. After each week, the nonspecific humoral (lysozyme, antiprotease and complement) response, cellular (myeloperoxidase content, production of reactive oxygen and nitrogen species) response and disease resistance against *A. hydrophila* were determined. Results indicated that the AqE of *E. alba* administered as feed supplement significantly enhanced most of the nonspecific immune parameters tested. Among the humoral responses, lysozyme activity significantly increased after feeding with AqE for 1, 2 or 3 weeks. No significant modulation was noticed in all the cellular responses tested after 3 weeks of feeding, while the '*reactive oxygen species*' (ROS) production and myeloperoxidase content showed significant enhancement after 1 week of feeding with *E. alba* AqE. When challenged with *A. hydrophila* after 1, 2 or 3 weeks of feeding, the percentage mortality was significantly reduced in the treated fish. The highest dose of 1 per cent gave better protection than the other doses with the relative percentage survival values of 64, 75 and 32 after feeding for 1, 2 and 3 weeks respectively. Finally, the results indicated that dietary intake of AqE of *E. alba* leaves enhances the nonspecific immune responses and disease resistance of *O. mossambicus* against *A. hydrophila* bacteria (Christybapita *et al.*, 2007).

Lonicera japonica herb has been known as an antiinflammatory agent, and is used widely for upper respiratory tract infections, diabetes mellitus and rheumatoid arthritis. *L. japonica* has significantly increased the blood neutrophil activity and promoted phagocytosis by the neutrophils in bovine at the correct concentration (Lee *et al.*, 1998). After challenge with *A. hydrophila,* survival of fish fed with the extracts of *Ganoderma lucidum* and *L. japonica* herbs was improved. The survival was further enhanced in the group fed with *Ganoderma* extract supplement and when both herbs were used together. This is the result of enhancement of some components of non-specific immune system of the fish by *Ganoderma* and a combination of *Ganoderma* and *Lonicera*. There is strong evidence that feeding glucans can modify the activity of the innate immune system of fish and increase the disease resistance in several fish species (Madhuri *et al.*, 2012e). The AqE of *G. lucidum* has promoted the phagocytosis by macrophages in mice immunosuppressed by an anticancer drug, cyclophosphamide. *G. lucidum* has stimulated the proliferation of lymphocytes induced by concanavalin A or lipopolysaccharide and influenced the gene expression of cytokines (Wang *et al.*, 2003). Many studies have shown that oral administration of chitin and yeast products (MacroGard, Vitastim and *Saccharomyces cerevisiae*) increased the phagocytic capability of cells in rainbow trout; extracellular activity was very high in fish fed with dietary glucan (Jeney *et al.*, 1997).

The non-specific immune effects of two Chinese herbs (*L. japonica* and *G. lucidum*) were determined on tilapia fishes. The herbal diets used were 1 per cent of *Lonicera*, 1 per cent of *Ganoderma* and a mixture of *Ganoderma* (0.5 per cent) and *Lonicera* (0.5 per cent). The diets were fed for 3 weeks. The respiratory burst activity of white blood cell (WBC), phagocytosis, plasma lysozyme, total protein and total immunoglobulin were monitored. Following 3 weeks after feeding, the fishes were infected with *A. hydrophila* and mortalities recorded. The study showed that feeding tilapia fishes with *Ganoderma* and *Lonicera* alone or in combination enhanced the phagocytosis by blood phagocytic cells during the whole experimental period and stimulated lysozyme activity after 2 weeks. Respiratory burst activity of phagocytic blood cells, total protein and total immunoglobulin in plasma were not enhanced. Both herbs when used alone or in combination increased the survival of fishes after challenge with *A. hydrophila*. The highest mortality (58 per cent) was observed in control fishes, followed with fishes fed with *Lonicera* extract (43 per cent) and fishes fed with *Ganoderma* (30 per cent). The lowest mortality (21 per cent) was observed when fishes were fed with a combination of these two medicinal plants. Hence, it can be concluded that the herbal extracts added to diets acted as immunostimulants, and appeared to improve the immune status and disease resistance of fish. It was also shown that *Astragalus* enhanced lysozyme activities in tilapia fishes during the whole period of the experiment when fed with low (0.1 per cent) and medium (0.5 per cent) doses. In case of fishes fed with *Scutellaria,* there was significant inhibition of extracellular superoxide anion production (Yin *et al.*, 2008).

Yin *et al.* (2008) further reported that the oral administration of ginger (*Zingiber officinale*) extract increases the phagocytic capability of cells in rainbow trout; while the extracts of four Chinese herbs (*Rheum officinale, Andrographis paniculata, Isatis indigotica* and *L. japonica*) increased the phagocytosis of WBCs of crucian carp. Ahilan *et al.* (2010) observed that the addition of *Phyllanthus niruri* and *Aloe vera* (Aloe) as herbal additives can positively enhance the growth performance of goldfish, *Carassius auratus*. The medicinal plants, *viz.*, ginger, nettle and mistletoe as an adjuvant therapy in rainbow trout through feed enhanced the phagocytosis, and cellular and humoral defense mechanisms against pathogens. The disease resistant of *Catla catla* fish was produced through immersion treatment of three herbs, *viz.*, A. *sativum* (garlic), *A. indica* (Neem) and *C. longa* (Haldi, turmeric) in spawn. Nargis *et al.* (2011) have seen the immunostimulant effects of the dietary intake of *A. sativum* and *Vitex negundo* extracts on the *L. rohita* (Indian major carp) fingerlings.

The immunostimulant effects of the dietary intake of various medicinal plant extracts on rainbow trout (*O. mykiss*), were investigated. The fishes were fed with diets containing AqE of mistletoe (*Viscum album*), nettle (*Urtica dioica*) and ginger (*Zingiber officinale*). The food containing lyophilized extracts of these plants as 0.1 per cent and 1 per cent was used at a rate of 2 per cent of body weight per day for 3 weeks. The plant materials tested for immunostimulatory food additives caused an enhanced extracellular respiratory burst activity ($P<0.001$). The fishes fed with a diet containing 1 per cent AqE of powdered ginger roots exhibited a significant non-specific immune response. Phagocytosis and extracellular burst activity of blood leukocytes (WBCs) were significantly higher in this group than those in the control group. All plant extracts added to fish diet increased the total protein level in plasma, except 0.1 per cent ginger. The highest level of plasma proteins was observed in the group fed with 1 per cent ginger extract containing feed. It was shown that in trout fed with nettle and mistletoe extracts the production of extracellular superoxide anion was of a similar level to that in the control fish (Dugenci *et al.*, 2003).

In *A. hydrophila* infected goldfish (*C. auratus*), fed with diets containing 100 and 200 mg kg^{-1} of mixed herbal extracts supplementation feeds, the WBC levels significantly increased ($P<0.05$). The RBC and haemoglobin (Hb) in goldfish significantly decreased ($P<0.05$) when fed with 100 and 200 mg kg^{-1} of mixed herbal extracts supplementation feeds; while it was restored near control when infected fish fed with 400 or 800 kg^{-1} of herbal extracts supplementation feeds. The haematocrit values declined significantly ($P<0.05$) in 100, 200 and 400 mg kg^{-1} of mixed herbal supplementation feeding groups on weeks 2 and 4 when compared to control group. The mean corpuscular volume, mean corpuscular Hb and mean corpuscular Hb concentration values almost significantly altered from the control values. The infected goldfish treated with 100 or 200 mg kg^{-1} of herbal supplementation feeds exhibited significant ($P<0.05$) decrease in total protein, glucose and cholesterol levels on week 1 to 4; whereas, it was restored when infected fish fed with 400 or 800 mg kg^{-1} of herbal supplementation feeds on week 4. In comparison to untreated control goldfish, the respiratory burst activity and phagocytic activity of blood cells was significantly enhanced in infected fish feeding with 200, 400 and 800 mg kg^{-1} of herbal supplementation feeds compared to the control. However, the infected fish fed with all the doses of mixed herbal supplementation feeds, the lysozyme activity was significantly enhanced throughout the experimental period. This study showed that the infected goldfish treated with 400 and 800 mg kg⁻¹ of herbal supplementation feeds preceding the challenge with live *A. hydrophila* had 30 per cent and 25 per cent mortality. Whilst, 100 and 200 mg kg⁻¹ of herbal supplementation feeds treated groups were found the mortalities of 50 per cent and 45 per cent, respectively. The results indicated that 400 or 800 mg kg⁻¹ of mixed herbal supplementation feeds restored the altered hematological parameters and triggered the innate immune system of goldfish against *A. hydrophila* (Harikrishnan *et al.*, 2010; Pandey *et al.*, 2012b).

Efficacy of dietary doses of *W. somnifera* root powder was evaluated on immunological parameters and disease resistance against *A. hydrophila* infection in *L. rohita* fingerlings. These fishes were fed with dry diet containing *W. somnifera* root powder for 42 days. Immunological parameters (*e.g.*, phagocytic activity, total immunoglobulin and lysozyme activity) of fishes were examined at 0, 14, 28 and 42 days of feeding. The fishes were challenged with *A. hydrophila* for 42 days post-feeding and the mortalities (per cent) were recorded over 14 days post-infection. Results showed that the fishes fed with *W. somnifera* root enhanced the phagocytic activity, total immunoglobulin level and lysozyme activity (p<0.05) compared with the control group. The survivability was higher in experimental diets than the control group. The dietary *W. somnifera* showed significantly (P<0.05) higher protection against *A. hydrophila* infection. The results suggested that the *W. somnifera* root powder has a stimulatory effect on immunological parameters and increases disease resistance in *L. rohita* fingerlings against *A. hydrophila* infection (Pandey *et al.*, 2012b; Sharma *et al.*, 2010).

Treatment of Fish Cancers by Herbal Drugs

Herbal Drugs for Treatment of Toxicity and Cancer in Fish

Billions of dollars (or rupees) are being spent every year all over the world, but no perfect remedy of cancer could be investigated so far. Although the herbal drugs (medicinal plants) have been with us for human therapy for millennia, there has been relatively little research on them to be used against fish diseases. The herbal drugs can be used not only as remedy of diseases, but they are also growth promoters, stress resistance boosters and preventatives of infections in the fish. Thus, the herbal drugs in disease management are gaining success, because they are cheaper and exhibit no or very less toxicity. They are rich in a wide variety of phytochemicals, such as tannins, alkaloids and flavonoids, which act against several diseases (Madhuri *et al.*, 2012e; Pandey and Madhuri, 2010a).

Further, the herbal drugs are known to have versatile immunomodulatory and antioxidant properties, leading to anticancer activities. The anticancer activity of the herbal drugs is due to presence of antioxidant phytoconstituents (*e.g.*, vitamins, carotenoids, terpenoids, flavonoids, polyphenols, flavonoids, enzymes, minerals, polysaccharides, alkaloids, saponins, lignins, xanthones and pigments). These antioxidants may prevent and cure cancer and other diseases by protecting cells from the *'free radicals'* damage (Pandey and Madhuri, 2010a). The herbal drugs can act as immunostimulants, conferring early activation to the non-specific defense mechanisms of fish and elevating the specific immune response. The herbal drugs have also been used in other countries for control of shrimp and fish diseases (Yin *et al.*, 2008). Many studies have proved that herbal additives enhanced the growth of fishes and protected from diseases. The non-specific immune system of fish is considered to be the first line of defense against invading pathogens (Ahilan *et al.*, 2010).

Anticancer Activity of Some Herbal Drugs in Fish

Christybapita *et al.* (2007) observed the immunostimulatory effect of the AqE of *E. alba* leaf (oral administration as feed supplement) in tilapia fish. On the basis of many studies, Yin *et al.* (2008) reported that oral administration of ginger extract increases phagocytic capability of cells in rainbow trout; while the extracts of four Chinese herbs (*R. officinale, A. paniculata, I. indigotica* and *L. japonica*) increased the phagocytosis of WBCs of carp. Ahilan *et al.* (2010) observed that the addition of *P. niruri* and *A. vera* as herbal additives can positively enhance the growth performance of goldfish. The medicinal plants, *viz.*, ginger, nettle and mistletoe as an adjuvant therapy in rainbow trout through feed enhanced the phagocytosis, and cellular and humoral defense mechanisms against pathogens. The disease resistant of *C. catla* fish was produced through immersion treatment of three herbs, *viz., A. sativum, A. indica* and *C. longa* in spawn. Nargis *et al.* (2011) have seen the immunostimulant effects of the dietary intake of *A. sativum* and *V. negundo* extracts on fingerlings of *L. rohita*.

The anticancer effect of *Solanum nigrum* (Makoi, black nightshade) herb was observed by Patel *et al.* (2009) in fish. They noted the significant cytotoxic effect of *S. nigrum* fruit methanolic extract (10-0.0196 mg/ml) on HeLa cell lines of fish. Efficacy of the polyphenol fraction of *Emblica officinalis* (*Phyllanthus emblica*, Amla, Indian gooseberry) fruit on the induction of apoptosis in mouse and human carcinoma cell lines, and its immunomodulatory effect on DEN induced liver tumours in rats was found. The polyphenol fraction of Amla induced the apoptosis in Dalton's lymphoma ascites and CeHa cell lines (Madhuri *et al.*, 2012e; Rajeshkumar *et al.*, 2003).

Chapter 5
Herbal Feed Supplements for Fish Health and Disease

Herbal Feed Supplements for Fish Illness

Herb-Feed Supplements for Health and Disease of Fish

Usefulness of herbs in treatment of diseases is attaining success, because the herbal treatment is cost effective, eco-friendly and has minimal side effects. Traditional herbal drugs may have the potential immunostimulation. Thus, use of herbs is an alternative to antibiotics in fish health management. Many reports have shown that herbal supplements enhance the growth of fishes and protect them from different diseases (Johnson and Banerji, 2007; Madhuri *et al.*, 2012g; Pandey, 2015). Inclusion of herbal additives in diets often provides cooperative action to various physiological functions. Synergistic effect of herbs has been reported in many fishes, including Japanese flounder and *Clarias gariepinus* (Madhuri and Pandey, 2014; Madhuri *et al.*, 2012g; Pandey, 2015; Pandey *et al.*, 2012c). The herbal growth promoters in the carp fish feeds showed beneficial effects. There was a significant difference between different herbal additives on the effect of growth rate in goldfish (Madhuri *et al.*, 2012g; Pandey, 2015).

The beneficial role of vitamins C and E have been extensively reported in fish nutrition, reproduction, growth and related indices. In addition, vitamins C and E are credited with modulating the stress response in fish. The beneficial utility of herbal growth promoters as an additive in the carp feed has been observed. There is a significant difference between different herbal additives on the effect of growth rate in goldfish. The non-specific immune system of fish is the first line of defense against invading pathogens. Neutrophils and phagocytes, lysozyme and complement are some important indices of non-specific immunity in fishes (Ahilan *et al.,* 2010; Madhuri *et al.,* 2012g Pandey *et al.,* 2012c).

Vitamins are the important essential nutrients for most animal species. Vitamin deficiencies in fish under aquaculture are known to produce biochemical dysfunction, leading to tissue and cellular level clinical manifestations. Many morphological and functional abnormalities have been reported in various fish species deprived of vitamins. Thus, the biological role played by vitamins C and E is very vital for the sustained growth and health of many living organisms as well as fish. Vitamin C is synthesized in animals from either D-glucose or D-galactose as part of the glucuronic acid pathway. Branching from L-gulonic acid, the biosynthetic pathway of vitamin C comprises three consecutive steps: first, the enzymatic lactonization of L-gulonic acid catalyzed by L-gulonolactone hydrolase; second, the oxidation of L-gulonolactone catalyzed by L-gulonolactone oxidase (GLO); and third, the spontaneous isomerization of 2-keto-L-gulonolactone leading to vitamin C. The general view is that the animals lacking GLO are unable to synthesize vitamin C and thus depend upon a dietary source of the vitamin. Among the fishes analyzed to date, only those retaining numerous ancestral characters, such as lamprey, shark, ray, lungfish and sturgeon have been shown to have GLO in the kidney, whereas teleost fish lack GLO activity (Madhuri *et al.,* 2012g; Pandey, 2015; Tatina *et al.,* 2010).

It has been documented that different factors are effective on the haematological and biochemical parameters of fishes, out of which the species, environmental condition, age, maturation and nutrition are very important. In fish, glucose concentration may vary greatly, depending on the physiological status of the animal. Normally, it is assumed that the nutritional state of a fish can affect the animal health and possibly the way they deal with stress. The stress response in fish is generally mediated by a neuro-endocrine response, which includes the release of stress hormones like cortisol and catecholamines into the circulatory system. These and possibly other hormones, elicit several compensatory physiological responses that help the fish to deal with the stressor. Glucose is one of the most important energy substrates used by fish to cope with physiological stress. Plasma glucose levels can increase, decrease, or keep constant under high plasma cortisol (Madhuri and Pandey, 2014; Madhuri *et al.,* 2012g; Pandey *et al.,* 2012c).

Fish cannot synthesize the essential amino acids (EAAs), which usually remain inadequate but they are required for the growth and development of body. The *'fish meal'* contains complete EAAs which meet the protein requirement of most fish species. Since the fish meal is costly as a feed ingredient, the use of nonconventional feed stuffs has been reported with good growth and better cost benefit values. The use of nonconventional feed stuffs of herb origin had been limited as a result of the presence of alkaloids, glycosides, oxalic acids, phytates, protease inhibitors, haematoglutinin, saponegin, momosine, cyanoglycosides and linamarin to mention a few despite their nutrient values and low cost implications. *'Non-conventional feed resources'* are feeds that are not usually common in the markets, and are not the traditional ingredients used for commercial fish feed production. They are very cheap, byproducts or waste products from agriculture, farm made feeds and processing industries, and are able to serve as a form of waste management in enhancing good sanitation. These include all types of feed stuffs from animals (silkworm, maggot, termite, grub, earthworm, snail, tadpoles, etc.), plant wastes (jack bean, cottonseed meal, soybean meal, cajanus, chaya, duckweed, maize bran, rice bran, palm kernel cake, groundnut cake, brewers waste, etc.), and wastes from animal sources and processing of food for human consumption like animal dung, offal, visceral, feathers, fish silage, bone, blood). All these can be recycled to improve their value if there are economically justifiable and technological means for converting them into usable products (Abowei and Ekubo, 2011; Pandey, 2013g and 2015).

The commercially prepared diets for channel catfish and salmonids have been developed for the specific nutritional requirements of these species, their production systems and their life stages. Some studies have been performed for tilapia production also. For all other species, including freshwater and marine ornamentals, nutritional management is based on a combination of application of knowledge generated for the species mentioned above and the experience of successful aquarists. The successful maintenance of difficult species is often influenced by the aquarist's success in obtaining or rearing specialized food items. For example, the sea horses and sea dragons (sygnathid family) have long, tubular mouth parts. These fishes are not physically capable of ingesting typical commercial fish feeds. Normally, the feeds for fry and fingerlings frequently exceed 50 per cent crude protein. As growth rate decreases with fish age, the protein levels in diets are decreased accordingly. The levels of protein on grow out diets often approach or exceed 40 per cent crude protein, while the maintenance diets may contain as little as 25 to 35 per cent. Many fishes require live food when they are hatched because their mouth parts are so small. In fish feeds, the fish meal should be a major protein source. There are essential amino and fatty acids that are present in fish meal but not present in tissue from terrestrial plants or animals. The low cost formulations in which fish meal has been eliminated and replaced by less expensive proteins from terrestrial sources (soybeans) are not recommended for fish. Fish meals and fishery byproducts have high lipid content, so the rancidity can be a problem if foods are not properly stored. In addition, fishes require long chain fatty acids (C20 and C22) which are not found in tissue from terrestrial organisms. There is a high oil content associated with carotenoid pigments, so vitamin E supplementation is recommended. Many fishes need dietary ascorbic acid (vitamin C). The ascorbic acid added to fish feed should be phosphorylated to stabilize the vitamin and increase storage time. In addition, vitamins A, D, E and B complex should be added to fish feeds. Concentration of vitamin E is generally inadequate, especially in diets which are high in fat (Abowei and Ekubo, 2011; Pandey, 2015).

Effects of Some Fish Herbal Supplements

Dietary vitamins were reported to have antibody enhancement effects in salmon fish. Disease resistance and humoral antibody production in rainbow trout fish was directly and positively related to the levels of vitamin C in the trout diet. In a 6-month-old white sturgeon fish fed a diet devoid of vitamin C, tissue total vitamin C concentrations were not decreased, suggesting that white sturgeon GLO produced adequate amounts of vitamin C to meet the fish needs at a stage of rapid growth while the dietary source was withheld (Madhuri *et al.*, 2012g; Pandey, 2015; Tatina *et al.*, 2010). Dietary vitamin C prevented the appearance of vitamin E deficiency signs in Atlantic salmon in a dose-dependent manner. Vitamin C deficiency also developed earlier in Atlantic salmon fed a diet high in vitamin E due to the accumulation of the vitamin E radical (tocopheroxyl) which is otherwise reduced by C vitamin. Hepatic vitamin E concentration decreased with increasing dietary C vitamin in hybrid striped bass. The hepatic vitamin E concentrations of normal lake sturgeon and vitamin E deficient yellow perch increased in fish fed with high dietary concentrations of C vitamin (Madhuri *et al.*, 2012g; Pandey, 2015; Pandey *et al.*, 2012c).

There are two interaction mechanisms between vitamins C and E: a synergistic simultaneous protection effect of the lipid and aqueous phases against oxidation and the action of vitamin C on vitamin E regeneration in the tissues. Data on growth, mortality, haematology and lipid oxidation in the liver revealed that vitamin C protects the fish against vitamin E deficiency (Pandey *et al.*, 2012c). It was also shown that supra-dietary levels of vitamins C and E may enhance antibody production and immune memory in juvenile milk fish to formalin-killed *Vibrio vulnificus* bacteria (Azad *et al.*, 2007).

Vitamins C and E function as biological antioxidants to protect cellular macromolecules (DNA, protein, lipids) and other antioxidant molecules from uncontrolled oxidation by *'oxygen free radicals'* (*'free radicals'*) during normal metabolism or under the conditions of oxidative challenge like infection, stress and pollution. This is why both vitamins have been known for protective actions against free redicals. Interactions between vitamins C and E have been proved in many *in vivo* studies since vitamin C spares vitamin E by regenerating it from tocopheroxyl radicals. Interaction between these two vitamins is also known to influence the beneficial effects they induce in cultured fish. Vitamin C/E sparing action in channel catfish was studied to explain the variability observed in its sensitivity to Vitamin E deficiency. So, due to their potential for interaction, dietary requirements for vitamins C and E are often considered together. A study was conducted on the starlet (a kind of fish) in order to characterize the different levels of dietary vitamin C (0, 100 and 400 mg kg^{-1} diet) and vitamin E (0, 100 and 400 mg kg^{-1} diet) on some of haematological and biochemical parameters. Results showed the highest number of leucocytes (WBC) in fish fed with diet containing 100 mg kg^{-1} vitamin E and 400 mg kg^{-1} vitamin C. The highest number of RBC was observed in different diets other than basal diet, which shows that the diets containing different levels of vitamin C and vitamin E have significant influence on RBC. However, different levels of vitamin C and E had not any significant influence on plasma protein and glucose. Fish fed with different levels of dietary vitamin C and E did not show any significant differences in cortisol and glucose values, which may be because the stress was avoided during experiment in all of the treatments. Different levels of vitamin C had no effect on the cholesterol amount. But diets containing excessive amounts of vitamin E and the basal diet without any vitamin supplement had the lowest significant amount of cholesterol which reveals that only an optimum amount of vitamin E can increase cholesterol and excessive amounts can lead to its decline. The dietary levels of vitamin C and vitamin E might have influenced some haematological and biochemical parameters of starlet fish. Similar results were obtained for channel catfish fed with different vitamin C and iron concentrations, for *S. aurata* fed with different vitamin C and E concentrations, for great sturgeon fed with different vitamin C levels, and for pirarucu fish fed with different vitamin C and E concentrations (Pandey, 2015; Tatina *et al.*, 2010).

In a study on the effect of different levels of vitamin E on immune response of grouper, it was shown that by increasing vitamin E in diet, the fish WBC count was increased accordingly. In another study on the effect of different levels of dietary vitamin C on some of the haematological parameters of great sturgeon, it was found that there were significant differences in WBC among the treatments. The high vitamins C and E concentrations could stimulate protein production in fish, suggesting an important role of both vitamins in the modulation of plasma proteins. The high levels of dietary vitamin E may decrease the amount of triglyceride and cholesterol in humans. It was explained that cortisol and glucose could increase in teleost exposed to stress. *Piaractus mesopotamicus* fish fed with diets containing 100 or 450 mg kg^{-1} vitamin E did not show a glycemic alteration compared to fish fed diets without this supplementation. However, the plasma glucose concentrations in pirarucu fish were elevated in 800 and 1200 mg vitamin E kg^{-1} treatments. So, it is not possible to confirm whether hyperglycemia is an advantage for these animals, since there is still a lack of standardization on vitamin E supplementation (Madhuri and Pandey, 2014; Pandey, 2015; Pandey *et al.*, 2012c; Tatina *et al.*, 2010).

The aqueous root extract of *Achyranthes aspera* plant was incorporated in the experimental diet of *L. rohita*. Feeding of fishes with this diet significantly (P<0.05) enhanced the serum anti-proteases level than fishes fed with control diet. *C. catla* were fed a diet containing seeds of *A. aspera* (0.5 per cent) for 4 weeks prior to and after ip injection with chicken erythrocytes (RBCs). The haemagglutination antibody titers, serum globulin levels, anti-trypsin activities and RNA/DNA ratio of spleen and kidney were found to be significantly higher in test group than the control group. These results confirm that *A. aspera* enhances the immunity of catla. In a similar study, *A. aspera* seed (0.5 per cent) was incorporated in the diet for *C. carpio* fish. After 4 weeks of feeding, fish were immunized with chicken RBC. Antigen-specific antibody response, total serum protein, serum albumin and globulin, lysozyme, serum a1-protease inhibitor and a2-macroglobulin and RNA/DNA ratio of spleen and kidney were significantly (P<0.05) higher, suggesting that the immune response of the fish was enhanced when fed with diet containing *A. aspera*. The *A. aspera* seed was incorporated in the diets (at 0.01 per cent, 0.1 per cent and 0.5 per cent) of *L. rohita* fingerlings. After 2 weeks, the fish were immunized with heat-killed *A. hydrophila* bacteria, and after a further 2 weeks, these fishes were experimentally infected with *A. hydrophila*. After 7 days, the superoxide anion production, serum bactericidal activity, lysozyme, alkaline phosphatase, serum protein and albumin:globulin ratio were enhanced (towards normal); while the serum glutamate oxaloacetate transaminase and serum glutamate pyruvate transaminase levels were decreased (towards normal) in fingerlings treated with *A. aspera* herb. The results indicated that this herb stimulates the immunity and increases resistance to infection in *L. rohita*. Furthermore, *A. aspera* after 4 weeks of its feeding to *C. catla*, significant (P<0.05) enhanced bovine serum albumin-specific antibody titers were seen. The efficiency of antigen clearance was also enhanced (Madhuri and Pandey, 2014; Madhuri *et al.*, 2012g; Pandey, 2015; Pandey *et al.*, 2012c). Goldfish fed with mixed herbal supplementation diets significantly restored the altered haematological (*viz.*, WBC, RBC, Hb, haematocrit value, mean corpuscular volume and mean corpuscular Hb concentration), biochemical (*viz.*, total protein, glucose and cholesterol) and immunological parameters, and triggered the innate immune system against *A. hydrophila* (Harikrishnan *et al.*, 2010).

Herbs promoted the cellular lipid and fatty acid utilization and protein accumulation resulting in good growth performance in *Pagrus major* fish (Ahilan *et al.*, 2010). The growth increase in *L. rohita* fish fed with herbal supplemented diet was due to improved food utilization and high protein synthesis (Johnson and Banerji, 2007). Among two different herbal additives, *P. niruri* fed group recorded higher specific growth rate, followed by *A. vera* fed group. The herbal incorporated feed fed fishes were challenged with *A. hydrophila* at the end of experimental period of 60 days. In *P. niruri* fed group, the highest survival rate of 75 per cent was noted at 1 per cent concentration, followed by 70 per cent at 1.5 per cent concentration. However, in case of *A. vera*, higher survival rate of 80 per cent was recorded at 1.5 per cent concentration, followed by 75 per cent at 1 per cent concentration (Ahilan *et al.*, 2010).

Fish Feed Ingredients of Plants Sources

Many fish farmers and ornamental fish hobbyists buy the bulk of their feed from commercial manufacturers. Small ornamental fish farms with an assortment of fish require small amounts of various diets with particular ingredients. Most feed mills will only produce custom formulations in quantities of more than one ton, and medicated feeds are usually sold in big bags. However, the small quantities of fish feeds can be made quite easily in the laboratory, classroom, or at home, with common herbal ingredients and with simple kitchen or laboratory equipment. The nutrients essential to fish are the same as those required by most other animals (Abowei and Ekubo, 2011; Pandey, 2013g and 2015).

The nutrients necessary to fish include water, proteins (amino acids), lipids (fats, oils, fatty acids), carbohydrates (sugars, starch), vitamins and minerals. In addition, pigments (carotenoids) are commonly added to the diet of salmonid and ornamental aquarium fishes to enhance their flesh and skin coloration, respectively. Soybean meal, legumes and wheat gluten are excellent plant sources of protein. Vegetable oils from canola, sunflower and linseed are common sources of lipids in fish feeds. Cooked carbohydrates from flours of corn, wheat or other breakfast cereals are relatively inexpensive sources of energy that may spare protein (which is more expensive) from being used as an energy source. The variety and amount of vitamins and minerals are so complex that they are usually prepared synthetically, and are available commercially as a balanced and premeasured mixture called *'vitamin or mineral premix'*. This premix is added to diet in generous amounts to ensure that adequate levels of vitamins and minerals are supplied to meet dietary requirements. A variety of natural and synthetic pigments or carotenoids are available to enhance the colouration in the flesh of salmonid fish, and the skin of freshwater and marine ornamental fish. The pigments most frequently used supply the red and yellow colours. The synthetically produced pigment, astaxanthin, is the most commonly used additive (100-400 mg/kg). Cyanobacteria (blue green algae, *e.g.*, *Spirulina*), palm oils and extracts from marigold, red peppers and *Phaffia* yeast are excellent plant sources of pigments. Another important ingredient in fish diets is a *'binding agent'*, which provides stability to the pellet and reduces leaching of nutrients into the water. Carbohydrates (starch, cellulose, pectin), and various other polysaccharides like extracts or derivatives from plants (gum arabic, locust bean) and sea weeds (agar, carageenin and other alginates) are popular binding agents. The preservatives like antimicrobials and antioxidants are often added to extend the shelf life of fish diets and reduce the rancidity of the fats. Vitamin E is an effective but expensive antioxidant which can be used in laboratory prepared formulations. The amino acids glycine and alanine, and the chemical betaine are also known to stimulate strong feeding behaviour in fish. Basically, the attractants enhance feed palatability and its intake. The fiber and ash (minerals) are a group of mixed materials found in most feed stuffs. In experimental diets, fiber is used as a filler, and ash as a source of calcium and phosphorus. In practical diets, both should be no higher than 8 to 12 per cent of the formulation. The live, frozen or dried algae are common feed stuffs used in ornamental fish diets. Fresh leafy or cooked green vegetables are often used in fish feeds. Although vegetables are composed mainly of water, they contain some ash, carbohydrates and certain vitamins. Kale, dandelion greens, parsley and turnip greens are examples of relatively nutritious vegetables. Many domestic and agricultural wastes, *e.g.*, corn bran, guinea corn bran, rice bran, wheat bran, palm kernel cake, groundnut cake, cotton seed cake, soybean, vegetable oil, palm oil, etc. are the main plant ingredients needed for compounding artificial fish diet to reduce the cost of fish production (Abowei and Ekubo, 2011).

Common Conventional and Unconventional Plant Feeds for Fish

Some common conventional feedstuffs from plant sources are groundnut cake, soybean meal, palm kernel meal, brewers dried yeast, brewers dried grain, maize and wheat offal. Groundnut cake contains about 45 per cent crude protein but lacks the essential amino acid, lysine. When moldy, it becomes poisonous due to the presence of mycotoxin called *'aflatoxin'*. Soybean meal is fast gaining increasing acceptability. It has balanced amino acids and can replace a substantial part of fish meal. Use of this feedstuff is, however, limited due to its high fat content and presence of trypsin inhibitor. Palm kernel meal contains a high quantity of crude fiber; crude protein is 17 per cent. Palm kernel meal is only useful when its crude fiber content is high. Brewers dried yeast is a byproduct of brewery industry. It contains sufficient quantity of crude protein but limited in amino acids methionine and cystine. Brewers dried grain is readily available and contains similar protein levels as palm kernel cake; crude fiber content is high and so this is in limited use. Maize is palatable and free from anti-nutritional factors. Its energy content is high. This limits the use in fish feed. Wheat offal has nutritional property similar to palm kernel meals. Thus, these two can be used interchangeably but scarcely together. Wheat offal is very scarce due to adverse government policy (Abowei and Ekubo, 2011).

The '*unconventional fish feeds*' are potential feed ingredients. They can be of animal or plant source. The plant sources of fish diets include leaf protein, leaf meal, aquatic macrophytes, cultivable pulses like mucuna bean, yam beans, bread beans, winged beans or any legume ornamental that can yield pods with seeds. Leaves contain diverse levels of protein, which can produce an inexhaustible and inexpensive source of nutrient for fish. Nutritionally valued leaves are cassava (*Manihot esculenta*), pawpaw (*Carica papaya*), pineapple (*Ananas comosus*), groundnut (*Arachis hypogea*), soybean (*Glycine max*) and plantain (*Musa paradisica*). Aquatic macrophytes are common aquatic plants found growing on water surface. These include rooted flowering plants like grasses and sedges that are commonly seen along the rim of freshwater bodies; rotted flowering plants with submerged leaves like ceratophyllum, and with floating leaves like water lilies (nymphaea); free floating plants like duckweed, water lettuce, water hyacinth and salvinia, a water fern. Water hyacinth are so wide spread that they constitute a menace to shipping and fishing activities but can be used as feed component for fish. A large number of legumes (unconventional pulses) are used as cover crops or ornamentals. They are not eaten by reasons of suspected content of toxic substances, *e.g.*, mucuna beans, broad beans, sword beans, winged beans, yam beans, etc. Their protein contents range from 18 to 20 per cent, fat 3 to 10 per cent and carbohydrate 50 to 60 per cent, making them easily gelatinisable. Toxic substances in them are hydrogen cyanide and trypsin inhibitors. They can be removed by heat during processing, which can be done by toasting (groundnut fashion), boiling, steam cooking or drying (Abowei and Ekubo, 2011; Pandey, 2015).

Fish Feed Preparation and Feeding Technology

Technology involved in rearing of live fish feeds is having a positive impact on larval rearing, a frequent bottleneck for commercialization of new species. Rotifers are the smallest live food regularly used for larval rearing. The newly hatched brine shrimps are larger but still quite small, and are commonly used in fish hatcheries. The cultured live foods can provide a source of high quality nutrition, but care must be taken to avoid perpetuation of infectious disease. Fish should be fed based on body weight percentage. For maintenance, 51 per cent body weight per day is sufficient, and fish should be probably fed at least 5 days per week. One feeding per day is plenty for most fishes. Rearing of young stock requires small meals fed more frequently. This is often accomplished using automatic feeders on commercial farms. Advances in the diagnosis and correction of nutritional diseases should be significant over the next few years as there seems to be a great deal of research activity in this area (Abowei and Ekubo, 2011; Pandey, 2015).

The fish feed are prepared, which may be dry with final moisture content of 6 to 10 per cent; semi-moist with 35 to 40 per cent water; or wet with 50 to 70 per cent water content. Most feeds used in intensive production systems or in home aquaria are commercially produced as dry feeds. Dry feeds may consist of simple loose mixtures of dry ingredients, *e.g.*, mash or meals, to more complex compressed pellets or granules. Pellets are often broken into smaller sizes known as '*crumbles*'. The pellets or granules can be made by cooking with steam or by extrusion. Depending on the feeding requirements of fish, pellets can be made to sink or float. Flakes are another form of dry food and a popular diet for aquarium fishes. They consist of a complex mixture of ingredients, including pigments. They are made into slurry, which is cooked and rolled over drums heated by steam. Semi-moist and wet feeds are made from single or mixed ingredients, *e.g.*, trash fish or cooked legumes, and can be shaped into cakes or balls. There is no single method for preparation of fish feed; however, most methods begin with the formation of a dough like mixture of ingredients. Dough is started with blends of dry ingredients which are finely ground and mixed. The dough is then kneaded and water is added to produce desired consistency for whatever fish is going to be fed. The same dough may be used to feed several types of fish, such as eels and small aquarium fish. Pelleting or rolling converts the dough into pellets or flakes, respectively. The amount of water, pressure, friction and heat greatly affects pellet and flake quality, *e.g.*, excess water in the mixture results in a soft pellet. Too little moisture and pellet will crumble. Proteins and especially vitamins are seriously affected by high temperatures. Thus, avoid storing diet ingredients at temperatures at or above 70°C and do not prepare dry feeds with water at temperature higher than 92°C (Abowei and Ekubo, 2011; Pandey, 2013g and 2015).

To make own fish feed, the tools are used primarily for chopping, weighing, measuring ingredients, and for blending, forming and drying the feed. Multipurpose kitchen shears, hand graters, a paring knife, a 5 inches serrated knife, a 6 to 8 inches narrow blade utility knife and a 10 inches chef knife for cutting, slicing and peeling can be used. A couple of plastic cutting boards protect the counter and facilitate the handling of raw ingredients. The heat resistant rubber spatulas, wooden and slotted spoons, long handled forks and tongs are very good for handling and mixing the ingredients. A basic mortar and pestle, electric blender, food processor or coffee grinders are very useful to chop or puree ingredients. A food mill and strainer like a colander or flour sifter help discard the coarse material and obtain fine food particles. For weighing and measuring the ingredients, dry and liquid measuring cups and spoons, and a food or laboratory bench scale are needed. Other utensils include plastic bowls (1½, 3, 5, and 8 quarts) for weighing and mixing the ingredients, a thermometer and a timer. A 3-quart saucepan and 10 inches stockpot are good for heating gelatins and cooking the raw foods like vegetables and starches. The ingredients and blends may be cooked in a small electric or gas burner. A few trivets to put under the hot pans will protect counters and table tops. The ingredients may be mixed by hand using a rotary beater or wire whisk; however, an electric mixer or food processor is more efficient. After mixing, a dough is formed which can be fashioned into different shapes. A pasta maker, food or meat grinder will extrude the dough into noodles or '*spaghetti*' of various diameters. As the noodles emerge from outside surface of die, they can be cut off with a knife to desired length or crumbled by hand, thus making pellets. A potato ricer also serves to extrude dough into noodles of same size. For making flakes, a traditional hand cranked or electric pasta maker will press out dough into thin sheets. The pellets or thin sheets can be placed on a cookie sheet and dried in a household oven on low heat or in a forced air oven. A small food dehydrator also performs the task quite well. To add extra oil and/or pigments to pellets, a handheld oil atomizer or sprayer can is useful. To separate pellets into different sizes, a set of sieves (*e.g.*, 0.5, 0.8, 1.0, 2.0 and 3.0 mm) is needed. Freezer bags serve to store the prepared feeds, and using a bag vacuum sealer will greatly extend shelf life of both ingredients and feed. The feed can be stored double bagged in the freezer but should be discarded after 6 months. One simple formulation, which is used traditionally to feed ornamental fish in ponds, consists of a mixture of 30 per cent ground and processed oats or wheat and 50 per cent of pellets from a commercial manufacturer. By weight, about 23 per cent of fish oil, and a 0.3 per cent vitamin and a 1 per cent mineral premix are added to mixture. This mixture is blended with water and can be formed into dough balls of different sizes (Abowei and Ekubo, 2011; Pandey, 2015).

With increased growth rate of stocked fish, the available natural foods in the pond become inadequate to support the fish population. So, there is need to supplement the natural food with artificial feeds for enhanced fish growth. The selection of fish for culture with supplementary diet depends on the crude protein requirement of the fish and the unit cost of the feed ingredient available in a given locality. This consideration is important because the fish feed accounts for 40 to 70 per cent of operating cost of a fish farm with an intensive management system. The feeding is one of the most important aspects of the fish culture. The range of type of food consumed by fish is greater than for other groups of vertebrates. Different technical methods are used for in administering feed to culture fish. Supplementary feeds are given to fish in addition to the natural food organisms in the water body. The supplementary feeds contain all the essential nutrients, *e.g.*, proteins, carbohydrates, lipids, vitamins and minerals required for fish growth. These feeds are introduced into the pond by broadcasting and spot feeding. In broadcasting, the feed is spread over the pond. Spreading is enhanced by drifting of the pond water. This method often contaminates the pond. Spot or spontaneous feeding involves placing the feed in a bag and tied on a spot. Alternatively, the feed is introduced at a particular time. The fish is fed (34 per cent body weight) twice daily, preferably morning and evening (Abowei and Ekubo, 2011; Pandey, 2013g and 2015).

Since fish feeds normally contain relatively high amounts of fish meal and/or fish oil, it is very susceptible to rancidity. Additionally, ascorbic acid is highly volatile, but critical to normal growth and development of most species of fish. Thus, fish feeds should be purchased frequently, ideally at least once a month and more frequently if possible. Fish feeds should be stored in a cool, dry place and should never be kept on hand for more than three months. Refrigeration of dry feeds is not recommended because of the high moisture content of that environment. The fish feeds which do not contain stabilized ascorbic acid are not recommended for fish. Pellets are typically the most complete diets. They are cooked, and, if marketed as a complete ration, the nutrition in each particle should be uniform. For larger animals, a very small pellet may be unacceptable. The semi-moist diets are soft and compact. Most of these are expensive, but they tend to be high quality diets and may be an excellent choice for some species of fish (Abowei and Ekubo, 2011; Pandey, 2015).

Herb-Feed Supplements act as Growth Promoter and Drugs for Fish

Herbal Drugs as Growth Promoters

Herbal drugs (herbs) are used not only against diseases but also as growth promoters, stress resistance boosters and preventatives of infections. Herbs can also act as immunostimulants, conferring the non-specific defense mechanisms of fish and elevating the specific immune response. Besides, the herbal drugs provide a cheaper source for treatment and greater accuracy than chemotherapeutic agents without causing toxicity (Madhuri *et al.*, 2012a and 2012g; Pandey *et al.*, 2012a).

Nutritional requirements of an animal are a fundamental aspect that depends on species, habitat and life cycle stage (Tatina *et al.*, 2010). Use of antibiotics and other chemotherapeutics for controlling diseases has been criticized for their negative impacts. The use of herbs as drugs in disease management is gaining success, because herbal drugs are cost effective, eco-friendly and have minimal side effects. Traditional herbal medicines seem to have the potential immunostimulation. Thus, the herbal drugs are alternatives to antibiotics for fish health. Many studies have proved that herbal additives enhanced the growth of fishes and protected them from diseases (Johnson and Banerji, 2007). Herbs are not only safe for consumers but also widely available throughout Asia, and also have a significant role in aquaculture (Ahilan *et al.*, 2010).

Vitamins C and E as herbal drugs or feed have beneficial role in fish nutrition, reproduction, growth and related indices. In addition, vitamins C and E are credited with modulating the stress response in fish. Vitamins are the important essential nutrients for most animal species. Vitamin deficiencies in fish under aquaculture are known to produce biochemical dysfunction leading to tissue and cellular level clinical manifestations. Several morphological and functional abnormalities have been reported in various fish species deprived of vitamins. Thus, it is clear that the biological role played by vitamins C and E as two important vitamins is very vital for the sustained growth and health of many living organisms as well as fish (Madhuri *et al.*, 2012g; Tatina *et al.*, 2010). Beneficial utility of herbal growth promoters as an additive in the carp feed has been noticed. There is a significant difference between different herbal additives on the effect of growth rate in goldfish. The non-specific immune system of fish is the first line of defense against invading pathogens. Neutrophils and phagocytes, lysozyme and complement are some important indices of non-specific immunity in fishes (Ahilan *et al.*, 2010; Madhuri *et al.*, 2012g).

Herbs promoted the cellular lipid and fatty acid utilization and protein accumulation resulting in good growth performance in *P. major* fish. The growth increase in *L. rohita* fish fed with herbal supplemented diet was due to improved food utilization and high protein synthesis (Madhuri *et al.*, 2012g).

Certain Herbal Feed Supplements

AquaImmu (AquaImmu Premix, also as ImmuPlus), a polyherbal drug (manufactured by Indian Herbs, Saharanpur, UP) contains the extracts of four Indian medicinal plants, *viz., Emblica officinalis* (Amla), *Ocimum sanctum* (Tulsi), *Tinospora cordifolia* (Giloe) and *Withania somnifera* (Ashwagandha). The recommended dosage of ImmuPlus Premix in fish, prawn and shrimp is 3 kg per ton (1000 kg) of feed. In these animals, AquaImmu is indicated for strong body defenses with optimum immune-status, to successfully meet the challenges of viral, bacterial and fungal infections; to improve and maintain the humoral and cell mediated immunity, and non-specific resistance to prevent the recurrence of infections and secondary infections; to optimize the vaccinial response with longer staying and higher antibody titres; to minimize the incidence of diseases and mortality thus increasing profits; and to maintain the health, livability, growth and performance at optimum levels (Madhuri and Pandey, 2014; Madhuri *et al.*, 2012g). ImmuPlus has been used to modulate the immune system of commercially important giant freshwater prawn, *Macrobrachium rosenbergii*. The prawns were fed with basal diet supplemented with ImmuPlus at the dose of 1g per kg of feed for 4 weeks. Results showed that the phenoloxidase activity, haemagglutination and lysozyme activities were significantly elevated in ImmuPlus fed prawn up to 3 weeks of feeding. Total protein level in ImmuPlus fed prawn raised up to 2nd week of feeding. Thus, the incorporation of ImmuPlus at the dose of 1g per kg of feed in the diet for 3 weeks may be beneficial in raising the immune status of prawn. Different experiments of AquaImmu (@ 1g/kg feed for 15 and 30 days) on fish (different stages of rohu) and prawn (*M. rosenbergii*), and suggested that the AquaImmu can be used as an immunopotentiator for healthy production of fish and prawns (Jayakumari *et al.*, 2004; Madhuri and Pandey, 2014; Madhuri *et al.*, 2012g).

AquaCE (AquaCE Premix) is anther herbal drug (Aqua herbal feed supplement prepared by Indian Herbs, Saharanpur), which contains the extracts of some Indian medicinal plants. AquaCE is fed to fish, prawns, shrimps and other aquatic species. It has all the desired effects for use as antistress, adaptogenic and antioxidant, and is safe and free from side effects generally associated with chemical products. It overcomes stress and minimizes the adverse effects of the stress response in fish, prawns, shrimps and other aqua species. The herbal ingredients of AquaCE are proven to have sustained antioxidant activity, more potent than activity of synthetic vitamin C and vitamin E combination. Thus while using AquaCE, there is no need for extra supplementation of synthetic these two vitamins. AquaCE also increases the adaptability of the body against stress by regulating the physiological and biochemical functions, and by increasing the non-specific resistance. The adaptogenic and antistress properties of AquaCE and its herbal components have been proved in many scientific studies against a variety of physical, chemical and biological stress conditions. It helps to maximize the production and profits in fish and aquaculture (Madhuri and Pandey, 2014; Madhuri *et al.*, 2012g). A trial was conducted to observe the beneficial effect of AquaCE against the bacteria (*Edwardsiella tarda*) on *L. rohita* fingerlings. Supplementation of AquaCE showed the lower mortality rate and total lymphocyte count. The globulin levels of fingerlings fed with AquaCE were recorded higher than the bacteria-challenged fingerlings. The experiment showed better natural resistance and survivability against virulent pathogenic bacteria (Mukherjee, 2008).

AmmoFree Premix herbal feed supplement has also been found effective in fish, prawns and shrimps suffering from ammonia poisoning. The gaseous ammonia is primarily released from the gills of these aquatic animals as a metabolic waste from protein breakdown and from the lesser secondary sources such as decomposition of uneaten feed and organic wastes by bacterial actions. Ammonia concentration is increased because of reduced algal population in the pond and due to reduced rate of assimilation of ammonia by plankton algae. Additionally, lower water temperature slows aerobic bacterial nitrification process whereby ammonia is converted to nitrates. High aquatic pH also severely accounts for the enhanced aqua-concentration of ammonia. Ammonia is extremely toxic and its toxicity is attributed mainly to its unionized form. Ammonia tends to block oxygen transfer from the gills to the blood and can cause both immediate and long-term gill damages. Similarly, AquaPro Premix has been developed by Indian Herbs for fish, prawns, shrimps and other aqua species. This herbal feed supplement is safe and has all the desired effects for use as performance enhancer and growth promoter. AquaPro Premix is a purely herbal drug containing natural herbs without any chemical transformation and is free from toxic or residual side effects, which are commonly associated with synthetic hormones, antibiotics and other chemical growth promoters, etc. It improves the intake and utilization of feed, promotes growth and thus ensures higher profits. Two other herbal feed supplements, *viz.*, MoultMore Premix and StickOn are available as growth promoter and medicines for fish, prawns and shrimps (Madhuri and Pandey, 2014; Madhuri *et al.*, 2012g).

Chapter 6

Toxicity to Fishes by
Drugs, Chemicals or Pollutants

Toxicity in Fish by Different Substances

Various Substances Causing Toxicity in Fish

Like humans and other species of animals, fish have also been found to be polluted with a large number of heavy metals, chemicals, drugs, infections or pollutants/contaminants. Many aquaculture chemicals are by their nature biocidal when released to the surrounding environment at toxic concentration. So, there is a potential for mortality of non-target organisms. Three classes of chemicals and their effects on non-target biota are: (1) carbaryl pesticide- mortality of non-target invertebrates, (2) organophosphate parasiticide- effects on nearby biota and (3) antibacterial residues- effects in aquatic sediments on the associated microbial community. The most common substances used in pond aquaculture are chemical fertilizers and liming materials. Less frequently used chemicals are oxidants, disinfectants, osmoregulators, algicides, coagulants, herbicides, probiotics and chemotherapeutants with agricultural industrial pollution. These compounds are quickly degraded or precipitated; they are not bioaccumulative and do not cause environmental perturbations in natural water receiving pond effluents. Most substances used in pond aquaculture to improve soil or water quality presented little or no risk to food safety. They may be biomagnified in the animal tissue and so consumers are at risk of intoxication with the chemicals (Madhuri *et al.,* 2012f).

The potential health hazards of handling industrial fish have been documented. Wet fish in a storage consumed oxygen produce poisonous gases as they are spoiled. Various noxious agents have been demonstrated in association with spoilage, including carbon dioxide, sulphur dioxide and ammonia. Cyanides are a further potential fatal complication of handling spoiled fish with noxious agent. The risk is especially high in industrial fishing because fish are stored in bulk without ice in closed spaces. The unintentionally added chemicals causing fish toxicity include organochlorine pesticides, PCBs and other persistent chemicals in feed, chemicals in construction materials, and metabolites and degradation products of intentionally added chemicals. Pesticides and piscicides (chemicals to kill fish) present a risk to human health and their use has to be carefully monitored (Hites *et al.*, 2004; Madhuri *et al.*, 2012f).

Heavy metals are considered most important form of aquatic pollution because of their toxicity and accumulation by marine organisms. The toxic effects of heavy metals, particularly As, Cd, Pb and Hg, have been broadly studied. Distribution of heavy metals varies between fish species, depending on age, development status and other physiological factors. Fish accumulate substantial concentrations of Hg in their tissues and thus can represent a major dietary source of this element for humans. Fish are single largest source of As and Hg for human. The primary sources of Hg contamination in humans are through eating fish. Metal contaminations in food, especially in marine products, have been broadly investigated (Emami Khansari *et al.*, 2005; Madhuri *et al.*, 2012e). The natural aquatic systems may extensively be contaminated with heavy metals released from domestic, industrial and other man-made activities. Fish are the inhabitants that cannot escape from detrimental effects of these pollutants. Fish are widely used to evaluate the health of aquatic ecosystems because pollutants build up in food chain and are responsible for adverse effect and death in aquatic systems. The studies carried out on various fishes have shown that heavy metals may alter the physiological activities and biochemical parameters both in tissues and in blood (Vinodhini and Narayanan, 2008; Madhuri *et al.*, 2012e).

Pesticides, disinfectants, antibiotics, chemotherapeutic agents and anesthetics are highly toxic. Use of large quantities of antibiotics in aquaculture has the potential to be detrimental to fish health, to the environment, wildlife and to human health. The Cu and Zn have been measured in sediments near aquaculture sites at high concentrations. These elements could be lethal to aquatic biota and persist in sediments. The Cu-based antifouling paints are applied to cages and nets to prevent the growth of attached marine organisms on them. The buildup of these organisms (epibiota) would reduce water flow through cages and decrease dissolved oxygen (Madhuri *et al.*, 2012f). The serious threats to aquaculture water come from herbicides used to control aquatic vegetation in fish ponds, runoff of pesticides, herbicides and fertilizers from fields adjoining aquaculture ponds; and aquifer contamination due to pollution of recharge water (Zakia *et al.*, 2012).

Effects of Certain Toxicants in Fish

Toxic effects of many heavy metals, chemicals, drugs, pollutants or contaminants have been studied extensively in fish by several researchers. A fatal case of methane and cyanide poisoning among a group of deep-sea trawler men was described. Methane is a well-known product of putrefaction and a potentially fatal load of cyanide can build up in a catch of fish as a secondary metabolite by bacteria-

Chromobacterium violaceum and *Pseudomonas* which acted as metabolic poison. A fatal case of inhalation of toxic fumes related to decaying fish was also noticed. The Cd and Pb levels in high dam lake water and fish (*T. nilotica*) are present as a result of the pollution that is attained from aquatic plants, sediments and gasoline containing Pb which leaked from fishery boats. The *T. nilotica* fish is used as a good bioassay indicator for lake pollution with Cd and Pb (Madhuri *et al.*, 2012f).

High levels of PCBs, dioxins and other contaminants have been reported in farmed salmon (Hites *et al.*, 2004). The fish caught in highly polluted water could cause cells of some kinds of cancers to multiply rapidly. The extracts from catfish caught from water high in sewage and industrial waste caused breast cancer cells to multiply. Fish contained substances that mimic the actions of oestrogen, the female hormone. As fish could concentrate some kinds of chemicals in water in their bodies, the results suggested that pharmaceutical oestrogen and xenoestrogenic chemical might be making their way into the region's waterways where the researchers exposed the extracts of catfish to oestrogen-responsive and oestrogen non-responsive human breast cancer cells. Excess ozonation is hazardous to humans as well as fish. Ozone gas that is released and comes in contact with live fish is also toxic. Ozone damages epithelial surfaces (*i.e.*, skin and gills) and will kill the fish (Madhuri and Pandey, 2014; Madhuri *et al.*, 2012f).

Some farmed fish have much higher body burden of natural and man-made toxic substances, *e.g.*, antibiotics, pesticides and persistent organic pollutants than wild fish. Farmed fish can have higher concentrations of certain toxic chemicals, especially manufactured chemicals than wild fish. The main sources of chemical contamination come from fish feed bioaccumulation and from location of aquaculture in contaminated areas. The former problems can potentially be minimized by changing the fish feed and by advisories on limiting the consumption of farmed fish, especially for susceptible individuals such as pregnant or nursing women (Cole *et al.*, 2009).

Problem of metal pollution is the most serious one. Industrial development in the developing and developed countries has resulted in heavy metals contamination of local water and affected the ecological balance. The metal pollution may damage marine organisms at the cellular level and possibly affect the ecological balance. Exposure and ingestion of polluted marine organisms as sea foods can cause health problems in people and animals, including neurological and reproductive problems. Chemicals of industrial effluents and products of ships and boats, such as heavy metals which find their way into different water systems can produce toxic effects in aquatic organisms. Petroleum products are one of the most relevant pollutants to aquatic ecotoxicology. Exposure to crude oil and derivatives can induce a variety of toxic symptoms in experimental animals. Petroleum hydrocarbons can act as a mediator in free radical generation in fish. The studies also revealed coastal pollution with several pollutants as well as heavy metals (Montaser *et al.*, 2010).

Liver cytotoxic alterations of adult medaka fish following short-term bath exposure (48 hr) to 500 mg per L DEN were studied for 3 to 21 days. Following DEN exposure, early changes in hepatocytes like nuclear lipid inclusions, nucleolar changes, decreased amounts of granular endoplasmic reticulum (GER), increased fractionation and steatosis of GER, proliferation of smooth endoplasmic reticulum and lysosomes, reduction in number and content of particulate lipoproteins and vitellogenin in golgi vesicles, and reduction in number and staining intensity of perioxisomcs were observed. Lesions like altered mitochondria and areas of spongiosis hepatis that developed at day 8, and infiltration of lymphocytes, granulocytes and macrophages at later stage were also seen (Bunton, 1996; Madhuri *et al.*, 2012e; Pandey *et al.*, 2014).

Heavy Metals Induced Toxicity in the Fish

Different Heavy Metals Causing Toxicity in Fish

The metal which has a relatively high density, and is toxic at low quantity is referred as *'heavy metal'*, *e.g.*, As, Pb, Hg, Cd, Cr, thallium (Tl), etc. Some *'trace elements'* are also known as heavy metals, *e.g.*, Cu, selenium (Se) and Zn. They are essential to maintain the body metabolism, but they are toxic at higher concentrations. The heavy metals can enter the bodies to a small extent via food, drinking water and air. The heavy metals concerned with environmental science mainly include Pb, Hg, Cd, Cr, Cu, Zn, Mn, Ni, silver (Ag), etc. (Pandey and Madhuri, 2014). In other words, the heavy metals are metallic elements which have a relatively high density, and are poisonous at low quantity. The excess quantities of heavy metals are detrimental as they destabilize the ecosystems because of their bioaccumulation in organisms, and elicit toxic effects on biota and even death in most living organisms (Gupta, 2013).

Due to formation of toxic soluble compounds, certain heavy metals become toxic. However, some metals are without any biological role or they are not needed by the body and become poisonous only in specific forms. Any amount of Pb can result to detrimental effect. *'Lighter metals'*, *e.g.*, beryllium can also be toxic in certain circumstances. Some *'essential elements'*, *e.g.*, Fe may also be toxic. Sometimes, action of essential elements can be changed by toxic metals, resulting into toxicity by interfering with metabolic process. So, most of the heavy metals are poisonous; while some metals are less toxic, *e.g.*, bismuth (Bi). Metalloids like As and polonium may also be toxic. Beside, both radiological and chemical toxicities can be induced by radioactive metals. Similarly, the metals with abnormal oxidation phase can also be poisonous, *e.g.*, Cr(III) is an essential trace element, while Cr(VI) exhibit carcinogenic effect. All heavy metals, in spite some of them are essential micronutrients, have toxic effects on living organisms via metabolic interference and mutagenesis. Such toxic effects include reduction in fitness, interference in reproduction leading to carcinoma and finally death. Both insoluble compounds and metallic forms of heavy metals generally have no toxic effect. The organometallic forms, *e.g.*, methyl Hg and tetraethyl Pb, can be highly toxic; while organometallic derivatives are less toxic, *e.g.*, cobaltocenium cation. The bioaccumulation of toxic metals can occur in the body and food chain. So, the toxic metals generally exhibit chronic toxicity. For example, the radioactive heavy metals like radium can imitate calcium (Ca) to be incorporated into the bone, but the similar health hazards can also be due to Pb or Hg. However, barium (Ba) and aluminium (Al) are exceptions as they can be quickly excreted by the kidneys. Industrialization has spoiled the environment by putting more and more concentrations of several metals. The heavy metals like Pb and Hg cause severe toxicity, as there are some historic cases, *e.g.*, Hg poisoning of waterways in Japan. The Hg toxicity (by methyl Hg) and Minamata disease exhibited significant neurotoxicity similar to *'Hunter Russell syndrome'*. Besides direct toxicity of heavy metals, the significant reduction in foetal growth and chronic effects should also be noted. Much of the basic research relies on studies in different animal species. Many heavy metals (including essential ones) are poisonous above their threshold levels. The heavy metals usually enter the body through respiration, ingestion and skin (Pandey and Madhuri, 2014).

At present, pollution has become a serious threat, and has brought hazards to the growing population as well as the earth/environment. The speedy urbanization and industrialization has led to increased disposal of pollutants like heavy metals, radio nuclides, and various types of organic and inorganic substances into the environment. Thus, the industrial wastes are the main source of metal pollution for aquatic organisms. Heavy metals constitute major pollutants in the environment. They are important pollutants for fish, because they are not eliminated from aquatic systems by natural methods like organic pollutants, and are enriched in mineral organic substances. The metal contaminants are mixed in aquatic system through smelting process, effluents, sewage and leaching of garbage which cause severe harm to aquatic system. Tannery industry has added pollutants to aquatic environment. The tannery waste waters continue to cause hazardous effects on the aquatic organisms as they also have endocrine disruption effects. A large number of chemicals are being used by the tanners during process, and thus discharge the toxic materials into waters. Due to this, the agricultural lands are also degraded. Uncontrolled release of tannery effluents has increased the health risks to different organisms (Praveena *et al.*, 2013).

Heavy metal pollution of aquatic environment has become a great concern. Because of toxicity and accumulative behaviour of heavy metals, they can make different changes in aquatic environment, such as species diversity. The heavy metals contents in aquatic environment have increased because of different activities like industrial, domestic and agricultural. With increasing heavy metals in the environment, these elements enter the biogeochemical cycle. The heavy metals can enter into water via drainage, atmosphere, soil erosion and all human activities by different ways. They can enter from contaminated water into fish body by different routes and accumulate in different organs. Some heavy metals, such as Fe, Cu, Zn and Mn are essential for biological systems like enzymatic activities; whereas, other heavy metals like Pb, Cd and Hg have no known important role in living organs and are toxic even in trace amounts. Essential metals must be taken up from water, food or sediment by fish for its normal metabolism; however, these metals can also have adverse and toxic effects at high concentration (Dobaradaran *et al.*, 2010).

Fish diversity of any regime has great significance in assessment of that zone reference to environment and pollution, as well as it contributes to the necessary information for fisheries. Many fishes may be the bioindicators of environmental pollutants also (Pandey, 2013h). However, in the conservation of fish diversity, it is essential to protect the fish from the environmental pollutant heavy metals, as the fish are most often contaminated by these pollutants. The As and inorganic As compounds, Cd compounds, Ni compounds, crystalline forms of silica, beryllium and its compounds have been said to be chemical carcinogens, resulting into the development of cancer in fish. The polluted marine organisms used as sea foods have caused health hazards, including neurological and reproductive disorders in both humans and animals. Nutrient pollutions cause outbreaks of fish diseases. Water contaminated with Hg can cause abnormal behaviour, slower growth and development, decreased reproduction, and death in fishes. *The persistent organic pollutants* can cause illness, deformities and deaths in fishes (Pandey and Madhuri, 2014; Pandey *et al.*, 2014).

Death in aquatic animals *may occur due to nutrient pollutions* (*e.g.*, nitrogen, phosphates, etc.) present in the toxic algae. Frog biodiversity and tadpole mass *can be declined by chemical contamination*. *Oil pollution* (as chemical contamination) can badly affect the development of marine organisms. It enhances the susceptibility to diseases and also affects reproduction. It can also cause gastro-intestinal irritation, and damage to liver, kidney and nervous system. *Deaths of* aquatic animals can also be due to high level of sodium chloride (NaCl) in waters. Manufacturing, farming, city septic systems, construction, automotive garages, laboratories, hospitals and other industries byproducts are highly toxic. These byproducts may be in different forms like liquid, solid or sludge which can contain chemicals, heavy metals, radiation, pathogens or other toxic materials. Items like batteries, used computer equipments and leftover paints or pesticides can also produce toxins. If these wastes are buried in the ground, or present in the stream runoff, in groundwater for drinking water, or in floodwaters, they are toxic to humans, animals, as well as plants. Toxic metals like Hg accumulate in aquatic system, and can be toxic to humans or animals when they eat fish. Toxic materials must be handled with precautions and be disposed off properly. More often, heavy metals are mixed in fertilizers. However, toxic levels of such metals are absorbed by some plants, ultimately consumed by humans resulting into detrimental negative effects on children (Pandey and Madhuri, 2014).

Poisoning of As is one of the most important toxicological hazards to animals which can occur due to arsenic trioxide, arsenic pentoxide, sodium and potassium arsenate, sodium and potassium arsenite, and lead or calcium arsenate. Drinking water containing more than 0.25 per cent As is potentially toxic. The As seems to prefer tissues rich in oxidative enzymes such as liver, kidney and intestine. The lethal doses of sodium arsenite in most species of animals are 1 to 25 mg per kg (Uppal, 2000). In a study on the spotted snakehead fish (*Channa punctatus*, Bloch), it was observed that when the high concentration (2 mM) of sodium arsenite (NaAsO) affected these fishes, they died within 2.5 hr. The chromosomal DNA of liver cells were fragmented which indicated that NaAsO might have caused death of those cells through apoptosis. The fishes were found particularly susceptible to As toxicity when they were continually exposed to it through gills and intake of As contaminated food (Ahmed *et al.*, 2008). Sensitive aquatic species were damaged by 19 to 48 μg As per L, 120 mg As per kg diet, or (in case of freshwater fish) tissue residues >1.3 mg per kg fresh weight (Madhuri *et al.*, 2012e).

High toxic levels of Hg (more than 0.05 ppm in a man weighing about 180 pounds) prevent the body's cells from transporting glucose, thus reducing energy available to body. This can produce convulsion, anorexia, tremor, swollen gum and behavioural disturbances in animals. In a case report, a human victim was killed by 300 ppm of Hg. The Hg *'burn'* was seen on the skin of the fishes. It is estimated that over 60,000 foetuses will suffer from methyl Hg toxicity in the utero from mothers eating swordfish, shark and tuna fish. LD_{50} values of Hg are as low as 1 mg per kg in small animals (Madhuri *et al.,* 2012e; Pandey and Madhuri, 2014).

Tuna, as a predator, is able to concentrate large amount of heavy metals, As, Hg, Cd, Pb and Sn was conducted in canned tuna fish (Emami Khansari *et al.,* 2005). In a study of Vinodhini and Narayanan (2008), *C. carpio* (common carp) fish was exposed to Cd, Cr, Ni and Pb at sublethal concentrations for periods of 32 days. The accumulation of heavy metal gradually increased in liver during the heavy metal exposure period. The order of heavy metal accumulation in the gills and liver was Cd>Pb>Ni>Cr and Pb>Cd>Ni>Cr, respectively. Similarly, in case of kidney and flesh tissues, the order was Pb>Cd>Cr>Ni and Pb>Cr>Cd>Ni, respectively. In all heavy metals, the bioaccumulation of Pb and Cd proportion was significantly increased in the tissues of *C. carpio*. Toxicity tests with rainbow trout fish confirmed that Cd is less toxic in hard water (96 hr LC_{50}=2.6 mg Cd/L) than in soft water (96 hr LC_{50}=1.3 mg Cd/L) (Madhuri *et al.,* 2012e).

Dobaradaran *et al.* (2010) determined the levels of Cd, Cu, Ni and Pb in muscle and skin of two important consumed fishes (Indo-Pacific king mackerel and Tigertooth croaker) in Bushehr Province in the Southwestern of Iran. The results showed that the highest and lowest contents of heavy metals in muscle and skin of both fish samples were related to Cu and Cd concentration, respectively, and heavy metal contents in both skin fish samples and muscle of Tigertooth croaker were found to decrease in sequence as Cu>Pb>Ni>Cd. In one more study (Montaser *et al.,* 2010), the histopathological lesions were reported in fishes exposed to Cd, Cu, Zn and Pb. Histopathological examination of liver revealed hepatocytes vacuolation, cellular swelling, nuclear degeneration and congestion of blood vessels. The gills exhibited secondary lamellar disorganization, rupture in lamellar epithelium and epithelial lifting (Madhuri *et al.,* 2012e).

The Cr is present in the tannery effluent, and is known to cause various ill effects. Such health hazards are dependent on the oxidation state of Cr. Its hexavalent form is toxic than the trivalent form. The haematological changes produced on the exposure to sublethal concentration ($1/10^{th}$ of LC_{50}/96 hr) of Cr have been observed in freshwater fish, *L. rohita* for 7 and 30 days, respectively. The decrease in haematological parameters suggested that the exposed fishes became anaemic due to exposure of Cr. Hence, this heavy metal is toxic which was discharged via the effluents into aquatic environments, and caused severe anaemia and alterations in haematological parameters in the *L. rohita* fish (Praveena *et al.,* 2013). The decreased levels of glycogen, protein and cholesterol have been observed after the administration of potassium dichromate in *L. rohita* (Pandey and Madhuri, 2014).

Heavy metal contamination may have devastating effects on the ecological balance of the recipient environment and a diversity of aquatic organisms. Because of the pollutants, a huge mortality occurs in different fish species (Madhuri *et al.,* 2012e).

Further, heavy metals are said to be the most important toxic pollutants for aquatic organisms. The As, Cd, Pb and Hg are tolerated at very low levels, and are highly toxic to humans. The Sn is very much used for studding in canning, but its toxicity is not much as of other heavy metals. The presence of metals in fish species depends on the age and development of fish, and other physiological factors. Fish are the single largest sources of As and Hg for human beings. The heavy metals released from domestic, industrial and other man-made activities can highly contaminate the aquatic systems. Such contaminations can seriously affect the ecological balance and diversity of aquatic species. For evaluation of health of aquatic systems, fish are widely used since pollutants present in food chain cause ill-effects and death of aquatic animals. Thus, the problem of metal pollution is one of the major health problems in the persons who eat sea foods. Due to metal pollution, cellular level damage has been observed, which possibly affect the ecological balance. The industries pour wastes containing Cr and large amount of chromates, dichromate, and other Cr compounds into the aquatic system. The Cr is widely used in industry, paints and metal platings as corrosion inhibitor, and its particulates enter the aquatic system through effluents discharged from tanneries, textiles, and electroplating, mining, dyeing, printing, photographic and pharmaceutical industries. The Cr exists primarily in the trivalent and hexavalent forms; but its hexavalent form predominates the trivalent form in the natural water. Contamination of natural water by Cr bearing industrial wastes is related to increased anthropogenic uses of this metal (Pandey and Madhuri, 2014; Pandey *et al.*, 2014).

Environmental and Health Risks by Heavy Metals

Heavy metals are accumulated in living organisms when they are taken up, and stored faster than they are broken down (metabolized) or excreted. They enter into water supply by industrial and consumer materials, or even from acidic rain breaking down soils and releasing heavy metals into streams, lakes, rivers and groundwater. The Pb, Hg and Cd are the most environmental heavy metals, but some other heavy metals can also badly affect the environment. *'Heavy metals toxicity'* is caused by different means; *e.g.*, from contamination of drinking-water (Pb pipes), high ambient air concentrations near emission sources, or from food chain. The heavy metals are poisonous since they bioaccumulate. *'Bioaccumulation'* means an increase in the level of a toxicant in a biological organism over time, compared to toxicant level in the environment. It is important to point out that the most of zoos which were once located on the outskirts of the cities and towns are now surrounded by human activities like vehicular traffic and industries. All these activities can cause heavy metal pollution, which may adversely affect the health and wellbeing of the wild animals housed in such protected areas (Gupta, 2013). Environmental and health risks caused by various pollutants heavy metals are as follows (Madhuri *et al.*, 2012e; Pandey and Madhuri, 2014; Pandey *et al.*, 2014):

Environmental and Health Risks by Arsenic

The As is commonly present in air, water, soil and all living tissues. It is at the 20th place found abundantly in the earth's crust, 14th in seawater and 12th in human body. It is a carcinogen, and can cause foetal death and malformations in many species of mammals. Most of the As compounds produced domestically are used in manufacture of agricultural products such as insecticides, herbicides, fungicides, algaecides, wood preservatives, and growth stimulants for plants and animals. Living resources are exposed to As by way of atmospheric emissions from smelters, coal-fired power plants and arsenical herbicide sprays; from water contaminated by mine tailings, smelter wastes and natural mineralization; and from diet, especially from consumption of marine biota. Generally, the inorganic As compounds are more toxic than organic compounds, and trivalent As (arsenites, As^{+3}) are more toxic than pentavalent As (arsenates, As^{+5}). The cancers of skin, lung, liver, lymph, nasal passage, kidney, bladder, prostate and haematopoietic systems of humans have been associated with inorganic arsenical toxicity. The As induced cancer risks have been especially prevalent among smelter workers, and in those engaged in the production and use of arsenical pesticides where atmospheric levels exceed 54.6 µg As per m^3. The inorganic forms of As can cause the symptoms like stomach and intestine irritation, skin disturbances, lung irritation and decreased WBC and RBC production. Very high exposure to inorganic As can also cause infertility, skin disturbances, declined resistance to infections, heart disruptions, brain damage and death. The acute oral LD_{50} values range from 10 to 300 mg per kg. The fatal doses of arsenicals are 2.5 to 33 mg per kg, orally in mammals. Susceptible species of mammals were adversely affected at chronic doses of 1 to 10 mg As per kg body weight, or 50 mg As per kg diet.

Environmental and Health Risks by Lead

Exposure to Pb can occur through ingestion and inhalation. Exposure of Pb can cause many effects depending on level and duration of Pb. The developing foetus and infant are more sensitive than the adult. Mostly, the bulk of Pb is received from food; however, other sources may be more important like water in areas with Pb piping and plumb solvent water, air near point of source emissions, soil, dust and paint flakes in old houses or contaminated land. In air, the Pb levels are brought in food through deposition of dust and rain containing metal on crops and soil. Eight broad categories of Pb use are: batteries; petrol additives; rolled and extruded products; alloys; pigments and compounds; cable sheathing; shot; and ammunition. In environment, the Pb comes from both natural and anthropogenic sources. The Pb exposure can be through drinking water, food, air, soil and dust from old paint. The Pb is among the most recycled non-ferrous metals, so its secondary production has grown steadily. The high levels of Pb may result in toxic effects in humans which in turn cause disruption of the biosynthesis of Hb, anaemia, high blood pressure, kidney damage, gastrointestinal tract (GIT) disorders, joints and reproductive/fertility problems, and acute or chronic damage to brain/nervous system.

Environmental and Health Risks by Mercury

Hg is not present naturally in living organisms. It is a toxic substance with no known function in biochemistry or physiology. It has complex and unusual chemical and physical properties. Degassing of earth's crust, emissions from volcanoes and evaporation from the natural bodies of water are the major natural sources of Hg. World-wide mining of metal leads to indirect discharges into atmosphere. The Hg is widely used in the industrial processes and in different products (*e.g.*, batteries, lamps and thermometers). It is also used in the dentistry as an amalgam for fillings and in the pharmaceutical industry. Prevalence of Hg in the environment leads to biomagnification in the food chain. The Hg is rapidly absorbed from coal fumes, contaminated seafood, pain, canvas and contaminated waters. This seriously affects the fish and food chain. Hence, the fish having high Hg content from the industrial pollution should not be eaten. Presence of Hg in oceans, and therefore in the ocean's fish, has detrimental effects on human beings. The Hg is mostly present in a relatively unreactive form as a gaseous element. The methyl (organic) forms of Hg are bioaccumulated over a million-fold and concentrated in the living beings, especially fish. Thus, these organic forms of Hg (monomethyl Hg and dimethyl Hg) are highly toxic than the inorganic forms due to ease of absorption into the human system. They cause neurotoxicological disorders. The inorganic Hg toxicity is associated with the tremors, gingivitis and/or minor psychological changes, together with the spontaneous abortion and congenital malformation in humans. The monomethyl Hg causes damage to brain and CNS, while foetal and postnatal exposures have given rise to abortion, congenital malformation and development changes in young children. The symptoms of Hg poisoning include kidney damage, disruption of the nervous system, damage to the brain functions, DNA and chromosomal damage, allergic reactions, sperm damage, birth defects and miscarriages.

Environmental and Health Risks by Cadmium

The Cd derives its toxicological properties from its chemical similarity to Zn (an essential micronutrient for plants, animals and human beings). The Cd once absorbed by an organism, it is present for many years (over decades for humans), though it is eventually excreted. This metal is produced as an inevitable by-product of Zn (or occasionally Pb) refining, since these metals occur naturally within the raw ore. But once collected, the Cd is relatively easy to recycle. The Cd is mostly used in Ni/Cd batteries, rechargeable or secondary power sources exhibiting high output, long life, low maintenance and high tolerance to physical and electrical stress. The coatings of Cd provide good corrosion resistance, particularly in high stress environments, such as the marine and aerospace applications where high safety or reliability is required; the coating is preferentially corroded if damaged. The Cd is also used as pigment, stabilizer for PVC, in alloys and electronic compounds. As an impurity, it is present in several products, including phosphate fertilizers, detergents and refined petroleum products. The average daily intake of the Cd for humans is 0.15 µg from air and 1 µg from water. The Cd if exposed for long time may cause kidney dysfunction. High exposure of Cd may cause obstructive pulmonary disease and lung cancer. The bone defects (osteomalacia and osteoporosis) have also been reported in humans and animals. Besides, the Cd can also cause increased blood pressure and myocardial disease in the animals. The exposure to Cd can damage kidney, CNS and immune system, and may also cause bone fractures and reproductive problems. The symptoms include stomachaches, diarrhoea and vomiting. The oral LD_{50} values of Cd in animals range from 63 to 1125 mg per kg.

Environmental and Health Risks by Selenium

Se is required in small amounts by the human beings and in larger quantities by the animals. It can damage to the nervous system, and cause fatigue and irritability. The Se accumulates in the living tissue, and its high contents in the fish and other animals can cause serious health problems in humans over a lifetime of overexposure. There may be hair and fingernail loss, damage to the kidneys and liver tissues, damage to circulatory tissues, and more severe damage to the nervous system.

Environmental and Health Risks by Antimony

The antimony (Sb) is used in compound, antimony trioxide (a flame retardant). It is also found in batteries, pigments, and ceramics and glass. Its high exposure for short duration can cause nausea, vomiting and diarrhoea. The long-term exposure of Sb can cause cancer in humans.

Environmental and Health Risks by Copper

In humans, Cu is essentially needed but in high doses, anaemia, liver and kidney damage, and stomach and intestinal irritation may occur. During Wilson's disease, it affects greatly. It is normally found in drinking water from Cu pipes and additives designed to control the algae growth.

Environmental and Health Risks by Chromium

The Cr has been reported to be used in metal alloys and pigments for paints, cement, paper, rubber and other materials. The low level Cr can irritate skin and can produce ulcer. Its chronic exposure can produce kidney and liver damage. The Cr can also damage to circulatory and nerve tissues. In aquatic animals, it is normally accumulated and can cause toxicity to eating fish. The hexavalent form of Cr is very toxic to man and animals. The Cr(VI) salts have many uses in divers industries and their indiscriminate introduction into aquatic ecosystem poses a serious threat to growth and survival of aquatic species, including fish population.

Environmental and Health Risks by Nickel

The Ni is needed in small amounts to produce RBCs, but it becomes slightly toxic in excess quantity. Its chronic exposure can cause decrease in body weight, heart and liver damage, and skin irritation. In aquatic animals, the Ni is accumulated but its presence is not magnified along the food chains.

Most Toxic Heavy Metals and their Toxic Effects

In cooperation with the US *'EPA'*, the *'Agency for Toxic Substances and Disease Registry'* (ATSDR) in Atlanta, Georgia (a part of the US Department of Health and Human Services) reported that in a *'Priority List for 2001'* called the *'Top 20 Hazardous Substances'*, As, Pb and Hg are at the 1st, 2nd and 3rd position, respectively in the list; while Cd is at the 7th place. Therefore, the heavy metals, *viz.*, As, Cd, Pb and Hg are considered most toxic to the humans, animals and environment. *Toxicities (detrimental effects) associated with these four* heavy metals, *and* the limits (*maximum daily dose* or exposure) of various elements/heavy metals for human and animals are respectively described in Tables 1 and 2 (Pandey and Madhuri, 2014; Pandey *et al.*, 2014).

Table 1: Toxicities Associated with Arsenic, Lead, Mercury and Cadmium

Heavy Metal	Toxicities
Arsenic (As)	Water-soluble inorganic As is readily absorbed from digestive system. Inorganic forms of As are particularly toxic. It causes irritation to lung, stomach and intestine, skin disturbances, and decreased formation of RBCs and WBCs. Very high concentrations of inorganic As can cause infertility, skin disturbances, decreased resistance to infections, heart disruptions, brain damage and death. The acute LD_{50} (oral) of As ranges from 10-300 mg/kg.
Lead (Pb)	It can enter the body through ingestion and inhalation. Its maximum allowable levels may be 5 µg/L (in bottled water) to set elemental impurities limit. It can cause disruption of biosynthesis of Hb, anaemia, high B.P., kidney damage, reproductive/fertility problems and brain or nervous system damage.
Mercury (Hg)	Its prevalence in environment can lead to biomagnification in food chain. The organic Hg, such as methyl Hg, is more toxic than inorganic Hg due to ease of absorption into human system. The toxicity of Hg include: kidney damage, disruption of nervous system, damage to brain, DNA and chromosomal damage, allergic reactions, sperm damage, birth defects and miscarriages. The LD_{50} of Hg is as low as 1 mg/kg in small animals.
Cadmium (Cd)	Cd is more readily absorbed through the lungs than the digestive system. It can damage kidneys, CNS and immune system. It can also cause bone fractures and reproductive problems. It can cause stomachaches, diarrhoea and vomiting. The LD_{50} (oral) of Cd in animals ranges from 63-1125 mg/kg.

Table 2: Limits (Maximum Daily Dose or Exposure) of Various Elements/Heavy Metals

Element/ Heavy Metal	Daily Dose (μg/day)		Element/ Heavy Metal	Daily Dose (μg/day)	
	Parenteral	*Oral/Topical/ Dermal/Mucosal*		*Parenteral*	*Oral/Topical/ Dermal/Mucosal*
Arsenic (Inorganic)	1.5	15	Nickel	25	250
Cadmium	0.5	5	Palladium	10	100
Lead	1	10	Platinum	10	100
Inorganic Mercury	1.5	15	Vanadium	25	250
Chromium	25	250	Osmium	10	100
Copper	250	2500	Rhodium	(combination not	(combination not
Manganese	250	2500	Ruthenium	to exceed)	to exceed)
Molybdenum	25	250	Iridium		

'Poison' is defined as any substance, which when absorbed into the body, will cause adverse or deleterious effects. Several metals and their compounds have been stated to be toxic to animals. The As, Cu, Pb, Hg and Cd have been reported to be the most toxic heavy metals. It is believed that many toxic metals exert their bad effects by distressing the enzyme systems of animals. Many of them bind to specific enzymes and proteins necessary for cellular function and thus compete with other substances essential for maintenance and the continued function of cells. So, poisons can also have the effect of inducing mineral deficiencies. In addition, many toxic appear to assist in the formation of the paramagnetic anion, superoxide (O_2^-), which itself is toxic and seems widely responsible for spontaneous cell death. Severe toxicity and bioaccumulation of several heavy metals pollution are related to aquaculture. By such aquacultures, the quality of fishes will be deteriorated, resulting into the detrimental health of fish eating population. Methyl Hg is thousand times more soluble in fats compared to water. The methyl Hg is concentrated in tissues of muscles, brain and CNS. The concentrations of Hg in fishes can be more than 10,000 to 100,000 times of the original levels in surrounding waters. Deposition of Hg is rapid but its depuration is slow. Less polluted shrimps are slow for depuration of Hg, but the polluted oysters depurate faster. The depuration of Hg is also found much slower in fish. Methyl Hg has its half-life of about 2 years in fish, but there is no significant health effect of general population by such Hg pollution. The levels of Hg for saltwater fish averaged 0.35 to 70.02 ppm and the levels of Se averaged 0.37 to 70.01 ppm. The Hg concentrations in bluefish are more than sufficient to produce toxic effects in mammals and birds eating fishes. The fishes bigger than 50 cm fork length have the average concentrations of higher than 0.3 ppm. This suggests that the pregnant women, children and others sensitive individuals should not eat such contaminated fishes. It means that fish consumption is the only chief source of methyl Hg for humans. The population that depends on daily fish intake can be more affected from long-term and high doses of methyl Hg and other organic pollutants. Likewise, high-end fish eaters are also severely exposed from Hg (Pandey *et al.*, 2014).

Heavy metals can have toxic effects on different organs. They can enter into water via drainage, atmosphere, soil erosion and all human activities by many ways. With increase of heavy metals in environment, they enter the biogeochemical cycle. They can enter from contaminated water into fish body by various routes and accumulate in organisms. These metals can be concentrated at different contents in organs of fish. The Cu and Zn have been measured in sediments near aquaculture sites at concentrations in excess of sediment quality guidelines. These elements could be lethal to aquatic biota and persist in sediments. In body, heavy metals enter through respiration, skin and intestinal absorption. Heavy metals of elemental forms are not fully absorbed; while organometallic forms are lipophillic and can soon enter through membranes, and even cross blood-brain barrier (BBB, the defense system of nervous system). Heavy metals after absorption into body can be widely distributed in different organs, including glands and CNS. Some of the heavy metals are called 'bone seekers' and they deposit into teeth and skeletal systems. The heavy metals then become toxic to enzyme system, and enhance free radical production and compete with essential elements which form metallo-enzyme complexes to compete with nutritional minerals absorption. Normally, all heavy metals cause toxicity to cells. After competing with nutritional minerals, they render them unavailable to body, leading to ill-health. For example, in a hair analysis sample, the Al displaces Ca so the latter is not available for bone formation, teeth and muscle (including heart muscle) function. Thus, these body structures become weak. The decrease concentration of Ca can lead to osteoporosis, heart disease, dental caries, periodontal disease, muscle cramping and colic. Other abnormalities include decrease concentrations of Zn, P (phosphorus), Mn, Fe and Mg (magnesium). The Al (having double whammy) increases Zn and Cu to secondary toxic levels (Pandey and Madhuri, 2014; Pandey *et al.*, 2014).

Toxicity of Mercury in the Fish

Exposure and Accumulation of Mercury in Fish

Heavy metals are considered the most important form of pollution of the aquatic environment because of their toxicity and accumulation by marine organisms. The natural aquatic systems may extensively be contaminated with heavy metals released from domestic, industrial and other man-made activities. With increasing heavy metals in the environment, these elements enter the biogeochemical cycle. The heavy metals can enter from contaminated water into fish body by different routes and accumulate in the organisms. These metals are concentrated at different contents in the organs of fish body. Fish may be more greatly affected by water pollution sources. They are highly exposed from the heavy metals, *e.g.*, Hg leading to severe toxicity, both in the fish and human beings (Pandey and Madhuri, 2014).

The Hg is widespread in the environment as a result of natural and anthropogenic releases. Everyone is exposed to small amounts of Hg over the course of a lifetime. Most atmospheric Hg is elemental Hg vapor and inorganic Hg. The Hg present in waters, soils, plants and animals is typically present in organic or inorganic forms. The organic Hg is primarily in the form of methyl Hg. The Hg is released into surface waters from natural weathering of rocks and soils, and from volcanic activity. It is also released from human action, *e.g.*, industrial activities, fossil fuel burning and disposal of consumer products (*i.e.*, Hg thermometers, fluorescent bulbs and dental amalgams). The global cycling of Hg via air deposition occurs when Hg evaporates from soils and surface waters to the atmosphere. From the atmosphere, the Hg is redistributed on land and surface waters; then absorbed by soils or sediments. Once the inorganic Hg is released into the environment, the bacteria convert it into organic Hg, which is the primary form that accumulates in fish and shellfish (Pandey *et al.*, 2012d).

The Hg vapor is emitted to the atmosphere through both natural and anthropogenic sources. The natural sources of Hg vapor include volcanoes, as well as rocks, soils and water surfaces. The Hg is also found naturally in cinnabar, the major ore for the production of Hg. Anthropogenic sources of Hg vapor include emissions from coal-burning power plants, municipal incinerators and through the recycling of automobiles (Clarkson and Magos, 2006). It is estimated that 50 to 70 per cent of the total emission of Hg to the environment is a result of human activity. About 1,000 twice as much, about 2,600 tons, is emitted from anthropogenic sources (Honda *et al.*, 2006).

The Hg emitted from all sources is cycled through the ecosystem. Once in the atmosphere, Hg vapor is slowly converted by oxidative processes to divalent Hg, which is then returned to the earth's surface by rainfall, where it accumulates in soils and in surface waters. Some of the Hg load is then converted back into Hg vapor and returned to atmosphere. However, another fraction of Hg load is washed into rivers, streams and eventually the ocean, where it accumulates in aquatic sediments. It is here that inorganic Hg is converted to methyl Hg by microorganisms living in the sediments by a process called *'methylation'* (Clarkson and Magos, 2006).

The methyl Hg then enters the food chain when it is absorbed by phytoplankton species. The phytoplanktons are eaten by plankton consumers, which then are eaten by larger fish and larger mammals. The methyl Hg accumulates in the tissues of fish and shellfish via a process called *'biomagnification'*, through which methyl Hg concentration increases as it moves up from one trophic level to the next. Within each organism, methyl Hg bioaccumulates as the organism consumes more and more organisms containing methyl Hg. Thus, smaller fish that are lower down in the food chain have lower concentrations of Hg in their tissues, while larger fish that are higher up in the food chain have higher concentrations. For example, sardines contain about 0.01 ppm of Hg, while sharks contain from 1 to 4 ppm. Fish with highest levels of Hg include sharks, swordfish and king mackerels. Large marine mammals like whales have levels similar to these fish. In aquatic food chain, methyl Hg biomagnifies as it is passed from lower to higher trophic levels through consumption of prey organisms. The fish at the top of food chain can biomagnify methyl Hg approximately 1 to 10 million times greater than concentrations in surrounding waters. About all of the Hg found in fish and other aquatic organisms are in the methyl Hg form. Long-lived predatory ocean fish may have increased methyl Hg content because of exposure to natural and industrial sources of Hg (Clarkson and Magos, 2006; Pandey *et al.*, 2012d).

Fish accumulate substantial concentrations of Hg in their tissues, and thus can represent a major dietary source of Hg for human. So, fish are the single largest sources of Hg for human. Primary source of Hg pollution in humans is through eating the fish (Emami Khansari *et al.*, 2005; Madhuri *et al.*, 2012f). Methyl Hg accumulation in seafood and fish products is a growing global concern that poses severe health risks to the public. While Hg occurs naturally, large amounts enter the environment from anthropogenic sources. Human exposure to Hg begins with production of many useful products. As only metal on the earth that can be found in a liquid form at room temperature, Hg and its compounds have many uses. Due to its special properties, including high density and high rate of thermal expansion, Hg is often used in barometers and thermometers. It can also be combined with other metals to create special alloys called *'amalgams'*. Gold and silver amalgams have been used in dentistry for fillings, and Sn amalgams are used to make mirrors. Hg can be found in many different lamps, including black lights, and is used in the industrial production of chloride and sodium hydroxide. Some Hg compounds are used as ingredients in skin creams, antiseptics, diuretics, fungicides, insecticides, and as a preservative in vaccines. The Hg compounds were even once used in the treatment of syphilis (Clarkson and Magos, 2006; Pandey *et al.*, 2012a).

The Hg is a naturally occurring heavy metal and a waste product of industries such as coal-burning power plants. Once Hg enters the water, it is consumed by microorganisms, which are eaten by small fish, and these, in turn, by bigger fish. At each step of food chain, Hg is retained in the muscle meat of fish, resulting in the highest concentrations of Hg in large, long-lived predatory fish, such as swordfish and shark (Pandey *et al.*, 2012d).

Mercury Toxicity in Fish Leading to Human

While some of the Hg compounds are fairly inert, many of them are extremely toxic. In USA, some of the Hg products containing have been banned, have limited use, or have special disposal requirements. These include dental fillings, vaccines, non-industrial thermometers, lamps, car starters and electronics. There are also many regulations regarding disposal of Hg wastes. Most organic Hg compounds are readily absorbed by ingestion and appear in lipid fraction of blood and brain tissue. Organic Hg readily crosses BBB and placenta. The foetal blood Hg levels are equal to or higher than maternal levels. Methyl Hg also appears in human milk. The organic Hg compounds are most toxic in the CNS, and may also affect the kidneys and immune system. The methyl Hg is toxic to the cerebral cortex and cerebellum in the developing brain and is a known teratogen. The Hg has a toxic effect on human nervous system, and can cause problems with learning, coordination and several other severe disorders, leading to death. The Hg is particularly dangerous for pregnant women (who can pass the Hg to the foetus through the placenta), breastfeeding women (who can pass the Hg to the baby through breast milk) and young children, whose nervous systems are developing. One form of Hg that is toxic and very harmful is elemental Hg. It is highly volatile and can easily be converted to Hg vapor, exposure to which can damage the nervous system, lungs and kidneys. For most people, exposure to Hg occurs when they eat fish or shellfish contaminated with methyl Hg. This is found in nearly all freshwater and marine fish (Pandey *et al.*, 2012d).

As body tries to rid itself of these toxins, gaseous Hg is oxidized to divalent Hg, which accumulates in kidney and can cause damage. Brain, kidney and lung are the target organs of elemental (gaseous) Hg. Brain and foetal brain are also the target organs for methyl Hg (Honda *et al.*, 2006). Most people are not exposed to inorganic Hg but rather absorb methyl Hg through consumption of fish and shellfish. Methyl Hg is easily absorbed in the digestive tract, where it forms a complex with amino acid cysteine. This new complex resembles a large neutral amino acid found in the body, methionine, and can more easily gain entry into cells. As with inorganic Hg, once in the bloodstream, methyl Hg will accumulate in brain and cause damage to CNS. Methyl Hg is naturally removed from the body over time. Eventually, this methyl Hg-cysteine complex is transported to liver, where it is secreted into bile, after which enzymes break the complex down into its amino acid and methyl Hg parts. Some of this methyl Hg then comes in contact with bacteria in the intestine and is broken down into inorganic Hg and carbon. Inorganic Hg is poorly absorbed in rest of the methyl Hg that does not interact with bacteria is reabsorbed by body and goes through the process again. It takes about 30 to 40 hours for methyl Hg to be distributed to the tissues of body. This cycle is the reason it takes so long to rid the body of Hg and how it can accumulate in the blood. It can take up to a year for Hg levels to drop significantly. Many adverse health effects are associated with the accumulation of Hg in body, though these vary depending on the amount of Hg one is exposed to, time of exposure, chemical form of Hg and age of the subject. Many toxic symptoms have been noted in people who consumed fish that were contaminated directly by methyl Hg from anthropogenic sources, not from methyl Hg that accumulated through natural methylation process (Clarkson and Magos, 2006; Pandey *et al.*, 2012d).

In the 1950's, one of the most severe incidents (known as the *'Minamata Bay Incident'*) of industrial pollution and Hg poisoning occurred in the small seaside town of Minamata, Japan. A local petrochemical and plastics company, Chisso Corporation, dumped an estimated 27 tons of methyl Hg into the Minamata Bay over a period of 37 years. The Hg was used as a catalyst in the production of acetaldehyde, a chemical employed in the production of plastics. The methyl Hg-contaminated waste water, a byproduct of the process, was pumped into the bay, creating a highly toxic environment that contaminated local fish. The residents of Minamata, who relied heavily on fish for food, were at risk of exposure to methyl Hg with every bite of fish they ate. The high contamination levels in the people of Minamata led to severe neurological damage and killed more than 900 people. An estimated 2 million people from the area suffered health problems or were left permanently disabled from the contamination (McCurry, 2006). This form of toxicity in humans is now called *'Minamata disease'*. Symptoms of this disease include sensory disorders of the four extremities, loss of feeling or numbness, cerebellar ataxia, tunnel vision or blindness, smell and hearing impairments, and disequilibrium syndrome. More serious cases lead to convulsions, seizures, paralysis and possibly death. In addition to the outbreak among the townspeople, congenital Minamata disease was observed in babies born to affected mothers. These babies demonstrated symptoms of cerebral palsy (Honda *et al.*, 2006).

Methyl Hg is 1,000 times more soluble in fat than in water, and it concentrates in muscle tissue, brain tissue and CNS. The Hg levels in fish may be in excess of 10,000 to 100,000 times the original concentration in surrounding waters. Its accumulation is fast, while depuration is extremely slow. The half-life of methyl Hg in fish is estimated at 2 years. The Hg levels for saltwater fish average 0.35 to 70.02 ppm (Burger, 2009; Madhuri *et al.*, 2012f). The methyl Hg content of fish varies by species and size of the fish as well as harvest location. Top 10 commercial fish species (*i.e.*, canned tuna, shrimp, pollock, salmon, cod, catfish, clams, flatfish, crabs and scallops) represent about 85 per cent of the seafood market and contain a mean Hg level of about 0.1 mg per g (Pandey *et al.*, 2012d).

Based on the growing body of evidence concerning the health issues of methyl Hg accumulation in the body, the EPA and *'Food and Drug Administration'* (FDA) of USA have issued advisories targeting consumption of fish for specific groups. Their advice to women who may become pregnant, pregnant women, nursing mothers and young children up to 6 years of age is to avoid certain types of fish high in methyl Hg, and limit the amount of fish consumed each week. Specifically, the EPA and FDA advise these groups not to eat shark, swordfish, king mackerel, or tilefish at all because they contain very high levels of mercury (>1 ppm). They also advise these groups to eat up to 12 ounces (or 2 average meals) a week of fish and shellfish that are low in Hg. Children should only eat 6 ounces of fish. The low Hg fish and shellfish include shrimp, canned light tuna, pollock, salmon and tilapia. *Albacore tuna* is a commonly eaten fish but contains moderate amount of Hg. The EPA and FDA advise eating only 6 ounces of albacore tuna a week. Also, if one exceeds the suggested amount of fish or shellfish in a week, simply cut back the amount consumed the next week or two. Lastly, the EPA and FDA advise the public to check for local advisories on fish caught from local lakes, rivers and streams. These fish may be more greatly affected by anthropogenic pollution sources. These guidelines are not aimed at adult men, or woman past child bearing age, but individuals concerned with possible exposure to Hg should follow them as well. The current action level of US FDA for Hg in fish tissue is 1 ppm (Pandey *et al.*, 2012d).

Levels of Hg in bluefish are high enough to cause potential adverse effects in sensitive birds and mammals that ate them and to provide a potential health risk to humans who consume them. Fish larger than 50 cm fork length with average levels of Hg above 0.3 ppm, suggest that eating them should be avoided by pregnant women, children and others who are at risk. Fish consumption is the only significant source of methyl Hg for public. Communities that relied on fish intake for daily nutrient sustenance may be at risk from chronic, high exposure to methyl Hg, as well as other persistent organic pollutants. Similarly, high-end fish consumers, whether recreational or subsistence, are at risk from Hg exposure (Burger, 2009; Madhuri *et al.*, 2012f).

Toxicities of Polychlorinated Biphenyls in Fishes

Polychlorinated Biphenyl Contamination to Fish

Like humans and other species of animals, fish are also polluted with a large number of pollutants/contaminants. Unintentionally added chemicals causing fish toxicity include organochlorine pesticides, polychlorinated biphenyls (PCBs) and other persistent chemicals in feed, chemicals in construction materials, and metabolites and degradation products of intentionally added chemicals. High levels of PCBs, dioxins and other contaminants have been reported in farmed salmon (Madhuri *et al.*, 2012f).

The PCBs are persistent environmental contaminants that are ubiquitous in the environment due to the intensive industrial use. These were used as commercial mixtures (Aroclor) that contain up to 209 different chlorinated biphenyl congeners, which are structurally similar compounds that vary in toxicity. A smaller subset of 50 to 60 congeners is commonly found in Aroclor mixtures. The PCBs' lipophilic character and resistance to metabolism enhance the concentration in the food web and exposure to humans and wild lives. Because PCBs do not burn easily and are good insulators, they were commonly used as lubricants and coolants in capacitors, transformers and other electrical equipments. Over the years, PCBs have been spilled, illegally disposed, and leaked into the environment from the transformers and other electrical equipment. The PCBs in the environment have decreased since 1970's but are still detectable in the air, water, soil, food and in our bodies. Lower chlorinated PCBs are more easily broken down in the environment, while adsorption of PCBs usually increases as chlorination of the compound increases. The highly chlorinated Aroclors resist both chemical and biological degradation in environment. The microbial degradation of highly chlorinated Aroclors to lower chlorinated biphenyls has been reported under the anaerobic conditions, as has the mineralization of biphenyl and lower chlorinated biphenyls by the aerobic microorganisms. Although these are slow processes, the volatilization and biodegradation are the major pathways of removal of PCBs from the water and soil, and volatilization is more significant for lower chlorinated congeners. Many PCB congeners persist in ambient air, water, marine sediments and soil at low levels throughout the world. In Puget Sound and other water bodies, the sediment-associated PCBs are accumulated in the bodies of aquatic organisms, which in turn are consumed by creatures higher in the food web. Fish, birds and mammals tend to accumulate certain congeners in their fatty tissues (Pandey *et al.*, 2012e).

PCBs can biomagnify in fresh and salt water ecosystems. Humans may be exposed to detectable quantities of PCBs when they eat fish, use fish oils in cooking, or consume meat, milk or cheese. Half life of PCBs in humans is estimated to be 2 to 6 years (Pandey *et al.*, 2012e). General population is exposed to PCBs by inhaling contaminated air and ingesting contaminated water and food. The dominant source of PCBs to humans is through consumption of meat, sea food and poultry. Of particular concern to this report is the exposure to citizens from consumption of fish. Some groups may consume greater amounts of fish than others, *e.g.*, Native Americans, Asian immigrant populations and sport anglers are three groups with high rates of sea food ingestion in the Puget Sound area (West *et al.*, 2001).

Fish is an essential component of the diets of certain local minority populations and Native Americans, who also eat fish that have higher levels of contaminants. However, knowledge of and adherence to health advisories for sport-caught fish differs by population. An epidemiologic study of Native American men found that 97 per cent were aware of the advisories regarding eating local fish. Knowledge of fish advisories may be low among minority populations, and these populations tend to consume fish that have higher levels of contaminants, *e.g.*, catfish and buffalo. PCB body burdens of fish-consuming populations in the Great Lakes basin who eat this fish are two-fold to four-fold higher than those in the overall US population. At-risk populations (*i.e.*, Native Americans, sport anglers, elderly pregnant women, and foetuses and nursing infants of mothers who consume contaminated Great Lakes fish) are continuously exposed to PCBs and other persistent substances like dioxins, chlorinated pesticides and Hg. Persons who ate Great Lakes sport fish for more than 15 years had 2 to 4 times more pollutants in their serum than non-fish eaters. The amount of fish consumed determines the level of exposure; fish consumption appears to be the major pathway of exposure; and a significant trend of increasing body burden is associated with increased fish consumption. Thus, primary pathway of exposure to persistent toxic substances (*e.g.*, PCBs) is from fish consumption. Persons who eat sport-caught fish consumed 2 to 3 times more fish than the overall US population. Women ate fish obtained from the Great Lakes during most of their reproductive years. Researchers found that reduction in breast milk PCB concentrations over time paralleled a corresponding decrease in the local fish consumption by the tribal women, and they concluded that this reduction in fish consumption might have resulted from the fish advisories that were issued recommending against the consumption of local fish by pregnant and nursing women. Their data indicated that 95 per cent of both men and women in the tribe are aware of these advisories and that 66 per cent of men and 40 per cent of women have changed fish consumption patterns (Pandey *et al.*, 2012e).

Toxicity of Polychlorinated Biphenyls in Fish

Toxic responses to PCBs are dermal toxicity, immunotoxicity, carcinogenicity, and adverse effects on reproduction, development and endocrine functions. Many epidemiological studies indicate that consumption of background levels of PCBs may cause slight but measurable impairments in physical growth and learning behaviour in children (Pandey *et al.*, 2012e). PCBs are probable human carcinogens and assigned them the cancer weight-of-evidence classification B_2 based on animal studies. Some information on pattern changes is available from studies in the Great Lakes. Early human effects from exposure to PCBs within the Great Lakes and St. Lawrence river basins have been observed (Humphrey *et al.*, 2000).

The PCB blood levels of frequent fish consumers have been significantly associated with non-Hodgkin's lymphoma. PCBs would contribute the majority of the non-cancer risks from Great Lakes fish consumption, although organochlorine pesticides could contribute some to the overall risk. There is a modest association of sport-caught fish consumption with the risk for conception failure in men. The women who had regularly eaten PCB contaminated fish during the preceding 7 years had their shorter menstrual cycles. The frequency of fish consumption and an index of a lifetime of PCB exposure appear to have a stronger relationship with menstrual cycle length than the number of years of fish consumption. In uterus, PCB exposure is associated with neurobehavioural deficits that can be assessed soon after birth (in newborn). The newborns of mothers in the high-exposure category exhibited a greater number of abnormal reflexes, less mature autonomic responses, and less attention to visual and auditory stimuli in comparison with newborns of low- or no-fish-consuming mothers, after adjustment for a variety of potentially confounding factors. The Michigan investigators found that the infants born to mothers who had eaten the greatest amount of PCB contaminated fish during pregnancy had more abnormally weak reflexes, greater motor immaturity and more startle responses, and less responsiveness to stimulation (Pandey *et al.*, 2012e).

A follow-up examination of children indicated that the neurodevelopmental deficits found during infancy and early childhood still persisted at age 11 years. The children had been exposed in utero through the consumption of PCB-contaminated fish by their mothers during the 6 years before and during pregnancy. After adjustment for many confounding factors, including maternal alcohol consumption, cigarette use, socioeconomic status, maternal age, parity of the mother and exposure to Pb and Hg, the results indicated that the most highly exposed children (based on maternal milk PCB concentration) were three times more likely than controls to have low full-scale verbal IQ scores, were twice as likely to lag behind at least 2 years in reading comprehension and have difficulty paying attention. These intellectual impairments were attributed to in utero exposure to PCBs and to related contaminants at concentrations slightly higher than those found in the overall population. Early nervous system dysfunction among adults exposed to PCBs and other persistent toxic substances has also been reported. This suggests that the nervous system alterations are associated with eating PCB-contaminated

fish, and that the deficit increases with consumption. The investigators have concluded that the Native Americans in their study tended to be higher consumers of fish, have elevated levels of Hg and PCBs in comparison with the overall population, and may be at higher risk for health effects (Pandey *et al.*, 2012e). The fish caught in highly polluted water can cause cells of some kinds of cancers to multiply rapidly. The extracts of catfish caught from water high in sewage and industrial waste cause breast cancer cells to multiply (Madhuri *et al.*, 2012f).

Cyanide Toxicity in the Fishes

Contamination of Cyanide to Fish

Contamination of inland and surface waters and land/soil due to the release of variety of chemicals may prove toxic to all classes of living organisms. One such dangerous and toxic chemical is *'cyanide'*. Although known for its potent toxic nature, cyanide has large applications in variety of industrial processes like metal mining (mainly Au and Ag), electroplating, steel, automobiles, carbonization, printed circuit board manufacturing and chemical industries, etc. Consequently, these industries emanate huge quantity of cyanide (HCN/CN^-) containing effluents. Wastewaters, especially generated from industries, *e.g.*, metal mining, electroplating, printed circuit board manufacturing, etc. are often toxic. Similarly, the precious heavy metals (*viz.*, Cu, Ni, Fe, Zn, Cd, Cr, Ag, Au, etc.) readily bind with free cyanide, resulting in the formation of metal-cyanide complexes (M_xCN- where M_x stands for metal moiety; shown in equation, below) of variable stability and toxicity (Prashanth *et al.*, 2011).

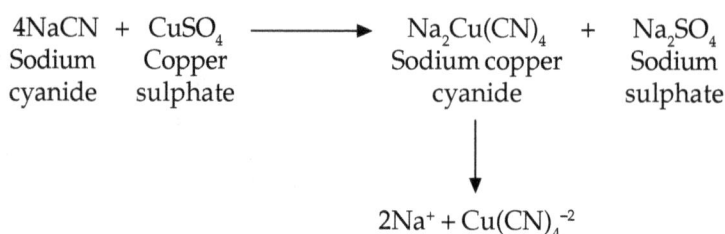

$$4NaCN + CuSO_4 \longrightarrow Na_2Cu(CN)_4 + Na_2SO_4$$

| Sodium cyanide | Copper sulphate | Sodium copper cyanide | Sodium sulphate |

$$\downarrow$$

$$2Na^+ + Cu(CN)_4^{-2}$$

The M_xCN form can occur in the waters and wastewaters in various forms, depending upon the concentration of metal and cyanide. For example, copper-cyanide (shown in equation) can occur in the water in various anionic forms like $Cu(CN)_2^-$, $Cu(CN)_3^{2-}$, $Cu(CN)_4^{2-}$, etc. The amount of effluent discharged daily by cyanide user industries is in the range of 200 to 1000 L for small-scale industries and 1 to 20 cubic metres or more for large-scale industries. Total cyanide and metal content of these effluents are in the range of 0.5 to 100 and 1 to 250 mg per L, respectively, as against to their permissible (discharge) limits of 0.2 and 0.01 to 5.0 mg per L, respectively. So, M_xCN containing effluents cannot be discharged in environment without giving proper treatment (Pandey, 2013d; Prashanth *et al.*, 2011).

Many physical-chemical methods are employed for the treatment, and the levels of total cyanides in discharged liquid wastes are brought down to 0.2 mg per L (200 µg/L). In the science of aquatic toxicology, fish play an important role in toxicity testing and hazard evaluation, as do the white rat and guinea pig in mammalian toxicology. Fish has been said to be very sensitive species to the cyanide toxicity (Prashanth *et al.*, 2011).

Cyanide Toxicity to Fish

Assessment of the toxicity of sodium cyanide with reference to aquatic biota, especially fish is crucial in establishing the toxicity evaluation. Sodium cyanide is known to be readily taken up by the aquatic organisms and bio-concentration factors range from 130 to 4,900 for various organisms (snail, daphnia and fish). In laboratory tests, sodium cyanide was reported to be highly toxic for aquatic organisms with median lethal concentration (LC_{50}) values ranging from 0.02 to 1 µg per L for newly hatched shrimp. The data indicates that sodium cyanide is toxic to *L. rohita*. The Indian major carp, *L. rohita* is an edible freshwater fish with great economic importance. Even 0.2 mg per L of cyanide concentration in aqueous systems is toxic for most of the life forms. There are reports that even cyanide concentrations as low as 0.01 to 0.1 mg per L are able to kill some sensitive animals species present in the waters (Prashanth *et al.*, 2011).

The LC_{50} of sodium cyanide (free cyanide) in *L. rohita* within 96 hr was found to be 33 µg per L. Behavioural changes when exposed to lethal concentration of sodium cyanide showed increased opercular movement, increased surface behaviour, loss of equilibrium, change in body colour, increased secretion of mucus, irregular swimming activity, rapid jerk movement, partial jerk and aggressiveness. The swimming behaviour was in a cork-screw palter, rotating along horizontal axis. In sublethal treatment, the schooling behaviour of the fish was slowly disrupted, and the ventilation rate was increased. The fish at 21st day of exposure exhibited balanced swimming and active feeding, and behaved in normal way. The O_2 consumption was decreased in lethal concentration (-22.64 to -70.13 per cent) but in sublethal concentration, decreased trend was improved and reached normal level at 21st day (-25.10 to -2.19 per cent). Alteration in O_2 consumption may be due to respiratory distress as a consequence of impairment in oxidative metabolism. Fish in sublethal concentration were found under stress, but that was not fatal. At the sublethal concentrations of sodium cyanide, the fish became irritable and hyper-excited. Jumping movements as well as restlessness were seen, and finally the fish turned upside down. Mucus secretion and loss of equilibrium were also noticed. They slowly became sluggish with short jerky movements, surfacing and gulping of air and erratic circular movements. Finally, they settled down at the bottom with loss of equilibrium and rolling of the body, and showed convulsions prior to death. The fish very often came to the surface in order to avoid toxic environment. Moreover, examination of the gill of dead fish revealed that the gill lamellae colour was changed from red to brown (Pandey, 2013d; Prashanth *et al.*, 2011).

The unusual behaviour of *L. rohita* in cyanide stress condition may be due to obstructed functions of neurotransmitters. The gill opercular movements increased initially to support enhanced physiological activities in stressful habitat and later decreased may be due to mucus accumulation of gill. The toxic stress of pesticides has direct bearing on tissue chemical compounds. The excessive secretion of mucus over the gills may inhibit the diffusion of oxygen during the process of gaseous exchange. It suggests that the sodium cyanide is not safe to non-target organisms like fishes (Pandey, 2013d; Prashanth *et al.*, 2011).

The migration of *L. rohita* fish to the bottom of the tank following the addition of sodium cyanide clearly indicates the avoidance behaviour of the fish, which was reported earlier in trout. The opecular movement of the fish ceases immediately following exposure to cyanide. The increase in opecular movement and corresponding increase in frequency of surfacing of fish clearly indicates that fish adaptively shifts towards aerial respiration (by obtaining atmospheric O_2 surfacing) and the fish tries to avoid contact with the cyanide through gill chamber. The increased ventilation rate by rapid, repeated opening and closing of mouth and opercular coverings accompanied by partially extended fins (caughing) in *L. rohita* by sodium cyanide toxicity could be due to clearance of the accumulated mucus debris in the gill region for proper breathing. The erratic swimming of the sodium cyanide treated fish indicated the loss of equilibrium, suggesting that the region in the brain associated with the maintenance of equilibrium, should have been affected (Prashanth and Patil, 2006; Prashanth *et al.*, 2011).

Many toxic symptoms induced by free cyanide were observed in the Indian major carp, *Cirrhinus mrigala* (Prashanth and Neelgund, 2008). The toxicity of cyanide was also evaluated in the freshwater fish, *C. catla* (Prashanth and Patil, 2006). The hyperexcitability of the fish invariably in the lethal exposure to free cyanide/pesticides may probably be due to the hindrance in the functioning of the acetyl cholinesterase enzyme (AChE) in relation to nervous system (Prashanth, 2003). It leads to accumulation of acetylcholine, which is likely to cause prolonged excitatory post synaptic potential. This may first lead to stimulation and later cause a block in the cholinergic system. Accumulation and increased secretion of mucus in the fish exposed to free cyanide may be adaptive responses perhaps providing additional protection against corrosive nature of the pesticide and to avoid the absorptions of the toxicant by the general body surface. The abnormal changes in the fish exposed to lethal concentration of free cyanide were time dependent. Since most fish breathe water in which they live, changes in the chemical properties thereof may be reflected in the animal's ventilator activity, particularly if the environment affects respiratory gas exchange. Toxicants from the environment mainly enter fish by means of their respiratory systems. A mechanism of toxicant uptake through gills probably occurs through pores by simple diffusion and is then absorbed through cell membranes. The decreased O_2 consumption of *L. rohita* under the lethal and sublethal concentrations of sodium cyanide indicates the sequence of the type of compensatory mechanism, if any, which operates within the animal to overcome the load of toxic stress. The decreased O_2 consumption by the whole fishes might be due to the respiratory distress as a consequence of the impairment of oxidative metabolism. Similar decline in the whole fish O_2 consumption in different species of fishes exposed to toxicants was seen. Gills are the major respiratory organs and all metabolic pathways depend upon the efficiency of the gill for their energy supply. Thus, damage to gill may cause a chain of destructive events, which ultimately lead to respiratory distress. Depletion in O_2 consumption was due to disorganization of respiratory function caused by rupture in the respiratory epithelium of gill. It is also due to the disturbance in mitochondrial integrity and decreased activities of some mitochondrial enzymes. In addition to gill damage, decrease in Hb and decrease in tissues respiration may also interfere with respiratory process resulting in respiratory failure. Surfacing phenomenon of fish observed under cyanide exposure might be due to hypoxic condition of fish (Prashanth *et al.*, 2011).

Excessive secretion of mucus over the gills due to cyanide stress might inhibit the diffusion of oxygen during the process of gaseous exchange. The coagulation of mucus on the gills caused demolition of various important processes like gas exchange, nitrogen excretion, salt balance and circulation of blood. The alternative reason for the decrease in the O_2 consumption would be due to the internal action of cyanide. This poison appears to alter the metabolic cycle at subcellular level. Greater decrease in the rate of O_2 consumption of the fish, exposed to lethal concentration than the sublethal concentration, might be due to the considerable damage to the gill structure, and also due to the greater precipitation of mucus upon gill filaments leading to the clogging of gills. Probably, suffocation imposed by the coagulated mucus film, and necrosis on the epithelial and inter lamellar cells of gills is one of the reasons for the death of fishes in lethal concentration. Greater decrease in the rate of O_2 consumption of the fish in the lethal concentration might also be due to the greater damage caused to RBCs, as evident by the drastic decrease in the number of these cells. Lowering of O_2 consumption of fish in the sublethal concentration of sodium cyanide may be mostly due to the lowering down of energy requirements, and if so, such lowering of maintenance energy requirement is to be considered adaptive and even strategic. This lowering of maintenance energy requirements may be achieved by reducing osmotic gradient through the lowering of electrolyte levels in the body fluids. Furthermore, there is evidence for a considerable metabolic reorganization and increased utilization of anaerobic metabolism in fishes exposed to sublethal concentration of cyanide (Pandey, 2013d; Prashanth *et al.*, 2011).

Chapter 7

Toxicity of Pesticides in Fish

Pesticides Toxicity to Fishes

Role of Pesticides

'Pesticides' are chemical substances that are used to kill, repel, or regulate the growth of biological organisms. This diverse group includes insecticides, herbicides, fungicides, nematicides, acaricides, rodenticides, avicides, wood preservatives and antifoulants (Madhuri *et al.*, 2012d; Pandey and Pandey, 2013; Satyavani *et al.*, 2011). The United Nation's *'Food and Agriculture Organization'* (FAO) defines *'obsolete pesticides'* as stocked pesticides that can no longer be used for their original purpose or any other purpose, and therefore require disposal. A broad definition of obsolete pesticides includes all pesticides (technical and formulations) which have passed their expiry date (2 years post manufacture date), all banned pesticides, damaged and degraded products, unusable formulations and packages, unidentified products, associated contaminated empty containers and old application equipment, other contaminated materials and equipment, buried pesticides and containers, and heavily contaminated soils (Madhuri *et al.*, 2012d; Satyavani *et al.*, 2011).

Roles of pesticides in promoting our health and economy are closely related. Although pesticides have brought tremendous benefits to mankind by increasing food production and controlling the vectors of man and animal diseases; but they have also posed potential health hazards to fish. Fish blood is highly susceptible to internal and external environment fluctuations because it is the vehicle for transport of such pollutants (Madhuri *et al.*, 2012d; Pandey and Pandey, 2013). Use of various pollutants, *e.g.*, pesticides, heavy metals, etc. in the aquatic environment and their deposition in the biotic system is known to cause several structural and functional changes in the biota. The toxic effect of pesticides to the blood of fishes has been studied by many workers. Obviously, due to acute effects of pesticides killing of fish may occur (Madhuri *et al.*, 2012d).

Pesticides (especially insecticides and herbicides) have become the part of environmental pollutants due to their widespread use in agriculture and disease control. However, many pesticides are immunotoxic and suppress cell-mediated immunity (Mukherjee *et al.*, 2006). Herbicides and insecticides used in agriculture may result in fish kills. Exposure to toxicants may not result in immediate fish kills, but may affect fish populations by decreasing fecundity (number of eggs produced), reducing the viability of sperm, eggs and larvae, decreasing life expectancy, and increasing the incidence of abnormalities and natural mortality. The US EPA estimated that >1.2 billion pounds of pesticides are applied to crops, forests, residential areas, public lands and aquatic areas in the USA each year. Release of these chemicals into environment creates a potential for unintended adverse health impacts to both humans and non-target wild lives (Laetz *et al.*, 2009; Madhuri *et al.*, 2012d; Pandey and Pandey, 2013).

Pesticides Contamination to Fish

The release of pesticide chemicals into the aquatic environment creates a potential for unintended adverse health impacts to both humans and non-target wildlife. Assessing the cumulative toxicity of pesticides has, therefore, been an enduring challenge for environmental health research as well as ecotoxicology. The cumulative toxicologic impact of pesticide contamination/pollution is of particular concern for salmon and steelhead fish population. Extensive surface water monitoring for pesticides has shown that current-use pesticides are frequently detected in the salmon-supporting river systems (Laetz *et al.*, 2009; Pandey and Pandey, 2013).

Pesticides have been reported to have negative ecological consequences on biota and the environment at large. On biota, the effect on the physiological processes, reproductive failure, mortality and instability in the ecosystem function are seen. Fish and other aquatic animals are exposed to pesticides in three primary ways: (a) dermally, *i.e.*, due to direct absorption through skin by swimming in pesticide-contaminated water, (b) breathing, *i.e.*, by direct uptake of pesticides through gills during respiration and (c) orally, *i.e.*, by drinking pesticide-contaminated water or feeding on pesticide contaminated prey. Continual contamination and pollution faced by *'Nigerian Coastal and Marine Environment'* from pesticides run-offs with resultant fish kills and human deaths (Chindah *et al.*, 2004). Fish kills occur when pesticides are improperly applied to or otherwise end up in water bodies through either misapplication or drift (Madhuri *et al.*, 2012d). Accidental killing of fish due to contamination of the aquatic environment remains among the most frequent poisoning cases (Helena and Zdenka, 2009).

It was noticed that dealers gave those pesticides to farmers, which were reportedly expired, and the farmers bought such pesticides. Unfortunately, farmers may not be aware of the source of the pesticides sold to them and unwittingly introduce dangerous chemicals into the environment (Satyavani *et al.*, 2011). When croplands are treated, some impacts of pesticides occur on non-target terrestrial and aquatic ecosystems, as well as on adjoining agro ecosystems (Surendra, 2010). Surface water can also be contaminated directly by pesticide spray drift- the travel and deposition of fine pesticide spray droplets away from their intended target when the spray is applied too close to water. Drift incidents can result in greater surface water contamination than either runoff or leaching. The expired pesticides may enter into water bodies due to usage of expired pesticides, purchased due to lack of knowledge about expiry and due to malpractices in trading. The expired pesticides, if present in ecosystem, may pose new issues of toxicological concern, which are not similar with unexpired product. India has at least 200 tons of obsolete pesticides. Thus, expired pesticides must be disposed carefully and care should be taken to avoid their discharge to water bodies and thus prevent loss to fish. Past salmon population extinctions and current declines have been caused by decades of habitat degradation, overharvest, hydropower operation and hatchery practices. Pesticide mixtures are common in aquatic environment, including lakes, rivers, streams and other surface waters that support aquatic life. Further, pesticides almost always occur in mixtures with other pesticides (Madhuri *et al.*, 2012d).

Various Pesticides Causing Toxicity to Fish

Some commonly used pesticides and their toxic concentrations (mg/L) in fish have been reported are: aldrin (0.01), aminotriazole (300), caebaryl (0.02), chlordane (0.004), DDT (0.003), diasinon (0.002), dieldrin (0.005), diquat (0.5), diuron (1.5), dursban (0.001), endrin (0.003), lindane (0.02), malathion (0.008), parathion (0.001), pyrethrum (0.01), rotenone (10.0), silvex (2.0), simazine (10.0), tepp (0.3), toxaphene (0.01) and zectran (0.1) (Madhuri *et al.*, 2012d). Common pesticides are shown in Figures 57 and 58.

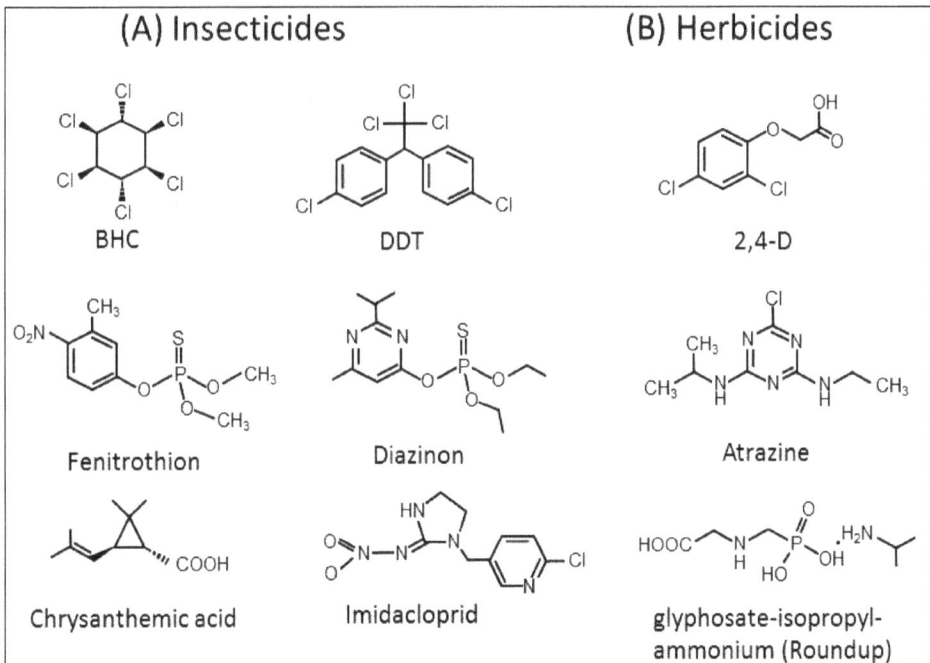

Figure 57: Common Pesticides.

Insecticide Class	Representative Structures	Other Major Products
Organophosphates - act by inhibition of acetylcholinesterase - major markets in Australia are the cotton, horticultural and livestock industries	*methyl-parathion:* 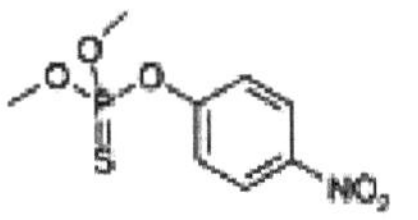 *chlorpyrifos:* 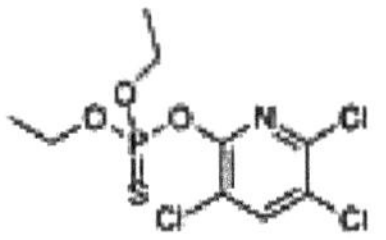 *malathion:* 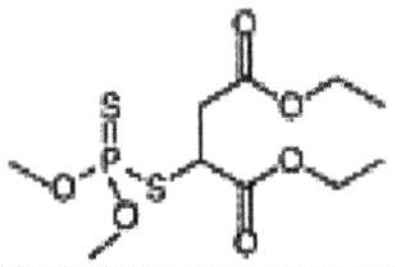	dimethoate omethoate azinphos-methyl diazinon chlorfenvinphos fenitrothion coumaphos dichlorvos monochrotophos
Synthetic Pyrethroids - act by binding to sodium channels - major market in Australia is horticulture	*permethrin:* 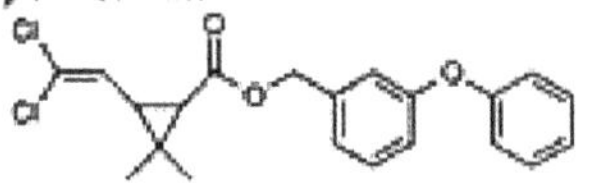 *cypermethrin:* 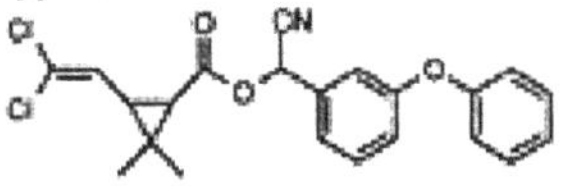	fenvalerate deltamethrin
Endosulfan - acts by binding to GABA receptors - major markets in Australia are cotton and horticulture	*endosulfan:* 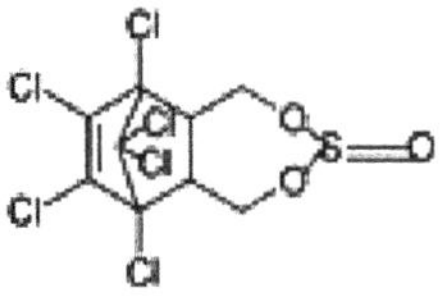	
Carbamates - act by inhibition of acetylcholinesterase - major market in Australia is horticulture	*carbofuran:* 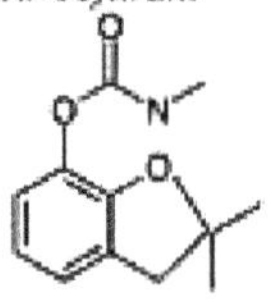 *carbaryl:* 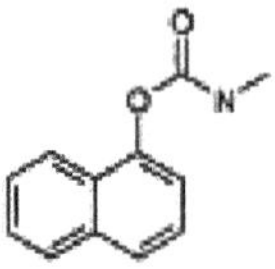 *pirimicarb:* 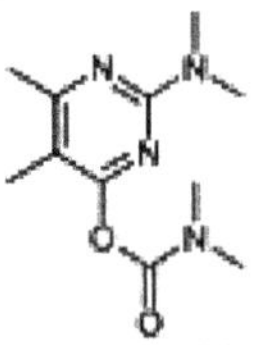 *methomyl:* 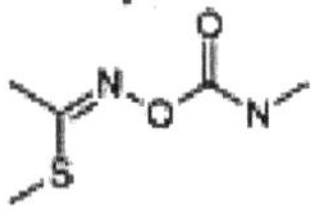	aldicarb thiodicarb

Figure 58: Common Pesticides.

DDT and alachlor pesticides can behave like endogenous oestrogen, and suppress apoptosis in estrogen receptor (ER)-positive human breast cancer. Some pesticides and related chemicals can act as carcinogens. Estrogenic endocrine disruptors or xenoestrogens are distributed in environment. Many pesticides like dieldrin, endosulphan, pentachlorophenol and sumithrin (pyrethroid insecticides), atrazine and DDT can act as endocrine disruptors (related to oestrogens). They can influence oestrogen functions, *e.g.*, adverse effects on release and excretion of hormones, disruption of regulatory feedback relationships between two endocrine organs, and modulation of non-genomic pathways. These xenobiotics adversely affect lymphocyte function, and increase oxidative stress and lipid peroxidation in various tissues (Mukherjee *et al.*, 2006; Pandey and Madhuri, 2010b). Dangers of flushing toxic chemicals into ecosystem through municipal sewer systems, one potentially devastating threat to wild fish populations comes from an unlikely source, oestrogen. Male fish exposed to oestrogen become feminized, producing egg protein normally synthesized by females. In female fish, oestrogen often retards normal sexual maturation, including egg production (Madhuri *et al.*, 2012d).

Toxic effect of pesticides on fish occurred largely during reproductive cycle (*i.e.*, at the time that the yolk sac was absorbed). Pesticide residues caused impairment in reproductive system and thyroid activity in aquatic fauna fish (Satyavani *et al.*, 2011). Toxicity of carbaryl and 1-naphthol (main degradation product of carbaryl, especially in the aquatic environment) to the aquatic organisms has been reported in the *L. rohita* carp (Madhuri *et al.*, 2012d). The cumulative toxicologic impacts of *'pesticide mixtures'* is of particular concern for salmon and steelhead populations that are currently listed as either threatened or endangered under the US *'Endangered Species Act'* (Laetz *et al.*, 2009).

In the years since enactment of the US *'Food Quality Protection Act'* (FQPA), the US EPA has identified several classes of pesticides that share a common mode of action. Among these are organophosphate (OP) and N-methyl carbamate insecticides. These two classes inhibit the acetyl cholinesterase (AChE) enzyme, thereby interfering with cholinergic neurotransmission in both humans and fish. Because anticholinesterase agents share a common mode of toxic action, National Academy of Sciences recommended a dose-additive approach to assessing risks to human infants and children. Dose addition (or for waterborne exposures to fish, concentration addition) assumes that the cumulative toxicity of the mixture can be estimated from the sum of the individual toxic potencies of each individual component chemical. This is how the US EPA currently assesses the potential toxicity of mixtures of OP and carbamate insecticides in the context of the FQPA. The assumption of dose addition or concentration addition for mixtures of anticholinesterase pesticides has also been extended to aquatic life. In salmon fish, concentration-additive inhibition of brain AChE activity by mixtures of OP and carbamate insecticides was demonstrated *in vitro*. However, the *in vivo* toxicity of anticholinesterase mixtures may deviate from concentration addition if the individual chemicals in a mixture interact via toxicokinetic or toxicodynamic processes to produce either antagonistic or synergistic effects. Each of these possible outcomes (antagonism, addition, or synergism) has potentially important implications for the current regulatory paradigm. Where the degree of synergism was severe (*e.g.*, for pairings of diazinon and malathion OPs), the enhanced AChE inhibition (*i.e.*, >90 per cent) corresponded to overt signs of anticholinesterase intoxication and death in fish. The data show that exposures to low, environmentally realistic concentrations of chlorpyrifos produced reductions in AChE activity that were closely correlated to reductions in swimming speed and feeding rates. Reductions in feeding are likely to lead to reductions in the size of exposed salmon at the time of their seaward migration, an end point that has been shown to be an important determinant of individual salmon survival. By reducing survival rates, sublethal inhibition of AChE in juvenile salmon could potentially reduce intrinsic productivity of salmon populations. Because mixtures of OP and carbamate insecticides produce dose-additive or synergistic AChE inhibition, they could magnify these population-scale effects. To define the extent to which OP and carbamate insecticides in mixtures interact, the juvenile coho salmon (*Oncorhynchus kisutch*) was exposed to all possible binary combinations of the OP insecticides like diazinon, malathion and chlorpyrifos and the carbamate insecticides like carbaryl and carbofuran (Laetz *et al.*, 2009).

Studies on toxicants, especially petroleum related hydrocarbon compounds indicated damage to epithelial cells of gill chamber, destruction of liver, possible nervous breakdown, failing organs and retarded physiological processes in fish body functions. The pesticide chloropyrifos (an organophosphate insecticide) showed deleterious effects on the survival, body and haematological processes in *Tilapia guineensis* fish (Chindah *et al.*, 2004). An increase in erythrocyte sedimentation rate (ESR) has been reported in *Clarias batrachus* fish after exposure to savin, and in *Heteropneustes fossilis* fish after exposure to alachlor and rogor. The increase in ESR might be due to increase in the concentration of fibrinogen which develops into fibrinogenemia after chlorpyrifos exposure (Madhuri *et al.*, 2012d; Malla *et al.*, 2009).

Atrazine is one of the most widely used herbicides, and is considered as a common terrestrial and aquatic contaminant. Atrazine at low concentrations causes kidney damage in chronic exposed rainbow trout. Endosulphan pesticide used in the cotton growing areas has been problematic over the last decade. In a case, pesticides were burned on a site which had subsequently been covered up with soil. Residues from incineration contaminated the surrounding soil, which local populations discovered was toxic to fish, and had been used by local populations to catch fish in the local river. The toxic soil would cause fish to die instantly, floating up to surface, where they would be caught and consumed or sold on the local market. High DDT and dieldrin residues in African fish at level which could potentially affect their reproduction, had chronic toxic and behavioural effects, and even drastically affected their population (Satyavani *et al.*, 2011).

Some farmers reportedly underestimated the toxicity of pesticides and thought that throwing of leftover pesticides in running water was their safe disposal. As pesticides degrade, they may leach into soil and water, or they may be windswept or volatilized reaching neighbouring, or far away areas (Surendra, 2010). Calculations revealed a significant relative toxicity of many active ingredients stored in each stockpile for fish. Some pesticide formulations were more toxic after shelf life expiry and some remained toxic but the variation may exist when compare with fresh respective formulation. Based on the LC_{50}, the formulations (*e.g.*, Dichlorvos, Fenvalerate, Lambda-cyhalothrin, Pretilachlor, Tebuconazole, Hexaconazole, Capta, etc.) having effective concentration (EC) may cause more toxicity in expired condition rather than other types of formulation. The altered pH may cause byproduct formation in expired pesticide which leads to acute toxicity to fish, *e.g.*, *L. rohita* (Madhuri *et al.*, 2012d; Satyavani *et al.*, 2011).

Organophosphates Toxicity in the Fishes

Organophosphate Pesticides and their Pollution

An *'organophosphate (OP) or phosphate ester'* is the general name for esters of phosphoric acid. Many of the most important biochemicals are organophosphates, including DNA and RNA as well as many cofactors that are essential for life. OPs are the basis of many insecticides, herbicides and nerve agents. The US EPA lists OPs as very highly acutely toxic to bees, wildlife and humans. The OPs are widely used as solvents, plasticizers and EP additives. They are widely employed both in natural and synthetic applications because of the ease with which organic groups can be linked together. The general formula of OPs is: OP(OR)(OR')(OR''), as shown in Figures 59–61.

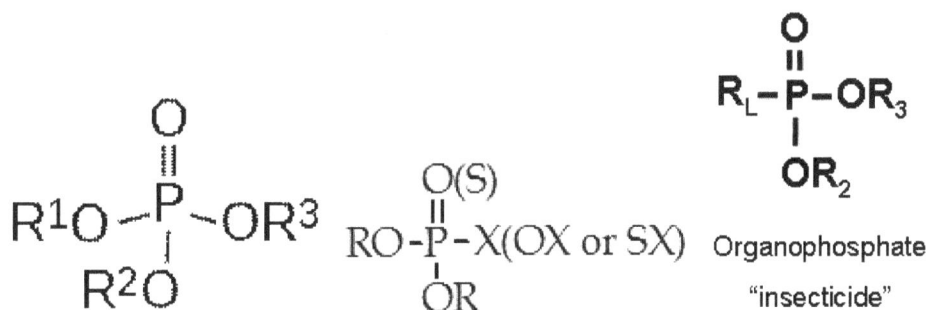

Figures 59–61: Organophosphate Pesticides in General.

Pesticides (herbicides, fungicides and insecticides) are widely used in the agriculture and industry around the world due to their high insecticidal activity. The presence of pesticide residues and metabolites in food, water and soil represents one of the major issues for environmental pollution. Pesticides are, in fact, among the most important environmental pollutants because of their increasing use in agriculture (Pandey and Pandey, 2013; Yue and Dai, 2011).

The OP poisoning most commonly results from exposure to insecticides or nerve agents. OPs are one of the most common causes of poisoning worldwide, and are frequently intentionally used in suicides in agrarian areas. There are around one million OP poisonings per year with several hundred thousand resulting in fatalities annually. The studies suggest a possible link to adverse effects in the neurobehavioral development of foetuses and children, even at very low levels of exposure.

Although the importance of multiple stressors is widely recognized in aquatic ecotoxicology (Eggen *et al.*, 2004), the pollution of pesticides continues to pose major challenges for natural resource agencies (Gilliom, 2007). These challenges include the data gaps that exist for many individual chemicals, experimental design difficulties (*e.g.*, near-insurmountable factorial complexity for large numbers of chemicals), poorly understood pathways for chemical interaction, potential differences in response among species and the need for more sophisticated statistical tools for analyzing complex data. The fish exposed to pesticides containing some of the most intensively used insecticides showed either concentration-additive or synergistic toxicity as well as unpredicted mortality. It implies that single-chemical assessments will systematically underestimate actual risks to fish species, or aquatic environment/ organisms in the watersheds where the pollution of OP pesticides occurs (Madhuri and Pandey, 2014).

Toxicity of Various Organophosphates in Fishes

Among the pesticides, OP and carbamate pesticides form an important class of toxic compounds; their toxicity is based on the inhibition of acetylcholinesterase (AChE). Three classes of pesticides have been problematic, *viz.*, organophosphates (OPs), carbamates and organochlorines. Organophosphate (OP) poisoning results from exposure to OPs, which cause the inhibition of AChE, leading to accumulation of acetylcholine (ACh) in the body, causing neurotoxicity. Presence of AChE enzyme in the insects, birds, fish and all mammals give this class of pesticides enormous toxicity towards unintended targets. Carbamate pesticides are also cholinesterase (ChE) inhibitors with a similar mechanism of action as OP pesticides (Pandey and Pandey, 2013; Yue and Dai, 2011).

Some examples of OP pesticides are diazinon, malathion and chlorpyrifos. Many OP insecticides are extremely hazardous to fish and wildlife. The OP insecticides can bio-concentrate in fish, frogs, tadpoles, and toads to levels that pose hazards to their predators. The OP insecticides are water soluble and metabolized quickly. They generally have short persistence (half-lives of days to months), and their residues do not pose long-term problems for aquatic animals (Laetz *et al.*, 2009; Madhuri and Pandey, 2014; Pandey and Pandey, 2013).

The OPs inhibit AChE, causing OP poisoning (Figure 62) by phosphorylating the serine hydroxyl residue on AChE, which inactivates AChE. The AChE is critical for nerve function, so the irreversible blockage of this enzyme, which causes ACh accumulation, results in muscle overstimulation. This causes disturbances across the cholinergic synapses and can only be reactivated very slowly, if at all. Paraoxonase

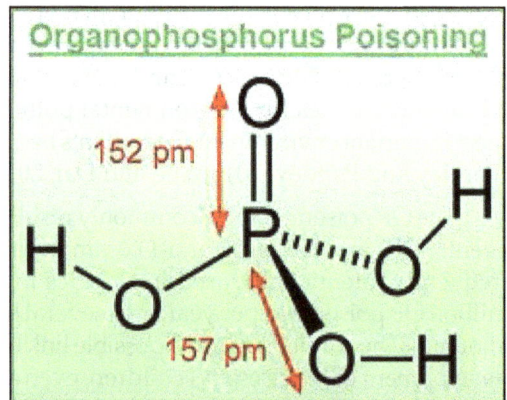

Figure 62: Organophosphate Poisoning.

(PON1) is a key enzyme involved in OP pesticides and has been found to be critical in determining an organism's sensitivity to OP exposure.

Chloropyrifos has deleterious effects on the survival, body and haematological processes in *T. guineensis* fish. *T. guineensis* exposed to sublethal concentration of chloropyrifos pesticide became anaemic and leucopenia due to reduction in erythrocytes (RBCs) and leucocytes (WBCs). This happened due to haemodilution resulting from impaired osmoregulation across the gill epithelium. The sublethal dose of chloropyrifos in the wild fish may cause growth lag, reproductive failure and/or death. *T. guineensis* exhibits behavioural responses to chemical stress both at acute and sublethal toxicity. Ecological importance of this is that damage to non-target species in environment and such attribute of organism could be effectively used as toxicity biosensor of chemical stress. The aquatic organisms (shell and fin fishes) that were in indirect continuous contact with the medium in addition to breathing and feeding had vulnerable to respiratory tract damage and other organs of body. Thus, chloropyrifos caused alterations in body functions and circulatory fluids, leading to mortality of Tilapine fish species (*T. guineensis*). A marked reduction in haemotological parameters of fish exposed to different concentrations of chloropyrifos was seen. This is an indication of the deleterious effects of the chemical pollutant to the body fluid (Chindah *et al.,* 2004). Chlorpyrifos kills insects by disrupting their nervous system, and is effective against a wide range of plant eating insects. Suspected effects of chlorpyrifos exposure include birth defects, increased rate of leukemia and immune system abnormalities. *C. punctatus* fishes were exposed with sublethal concentration of chlorpyrifos for 24, 48, 72 and 96 hr. A significant increase in ESR was seen, suggesting the toxic effect of chlorpyrifos after both acute and subacute exposure (Madhuri *et al.,* 2012d; Malla *et al.,* 2009).

The brain acetyl cholinesterase (AChE) inhibition was determined in juvenile coho salmon fish (*O. kisutch*) exposed to sublethal concentrations of OPs (*e.g.,* diazinon, malathion and chlorpyrifos), as well as the carbamates (*e.g.,* carbaryl and carbofuran). The investigators observed addition and synergism of these pesticides, with a greater degree of synergism at higher exposure concentrations. Many combinations of OPs were lethal at concentrations which were sublethal in single-chemical trials. It was concluded that the single-chemical risk assessments are likely to underestimate the impacts of these pesticides on salmon in river systems where the pesticides mixture occur. Moreover, the mixtures of pesticides which have been commonly reported in salmon habitats may pose a more important challenge for species recovery than previously anticipated. Based on a default assumption of dose addition, the all the above five pesticides were combined in all possible pairings to yield predicted AChE inhibitions of 10 per cent, 29 per cent and 50 per cent in the brains of exposed coho salmon. All binary pesticide combinations produced toxicity that was either additive or synergistic, with the frequency of synergism increasing at higher exposure concentrations. In all cases, the joint toxicity from the paired exposures resulted in AChE activities that were significantly lower than carrier controls (one-way ANOVA; $p < 0.05$). Coho exposed to combinations of diazinon and malathion (1.0 and 0.4 EC_{50}), as well as chlorpyrifos and malathion (1.0 EC_{50}), had the lowest measured AChE activities. Many fish species died after high rates of acute brain AChE inhibition

(>70-90 per cent). As expected, 100 per cent mortality was observed within the first 24 hr among coho exposed to the above pesticide combinations. Fish exposed to these OP mixtures also showed qualitative signs of anticholinesterase toxicity, including loss of equilibrium, rapid gilling, altered startle response and increased mucus production. The biochemical indicators of synergism (greater than additive AChE inhibition) were consistent with classical signs of anticholinesterase intoxication and death for salmon exposed to mixtures of organophosphate pesticides. Many insecticide combinations produced additive toxicity at low, environmentally relevant concentrations ($0.1\ EC_{50}$). Moreover, certain combinations showed a clear pattern of synergism even at these relatively low levels. For example, diazinon and chlorpyrifos were synergistic when combined at 7.3 µg/L and 0.1 µg/L, respectively. The pairing of diazinon (7.3 µg/L) with malathion (3.7 µg/L) produced severe (>90 per cent) AChE inhibition as well as classical signs of anticholinesterase poisoning. Thus, for some chemical combinations, synergism is likely to occur at exposure concentrations below the lowest levels used in the study. Although more work is needed to determine the lower bounds for pesticide interactions, this study indicates that synergism is likely to occur at concentrations that have been directly measured in habitats supporting threatened and endangered salmonids (Laetz *et al.*, 2009).

Carboxylesterases (CaEs) play an important role in the detoxification of many pesticides, including the OPs and carbamates insecticides, via hydrolysis. CaEs may also functionally protect AChE from insecticide toxicity by direct binding and sequestration, thereby preventing or delaying interaction between the insecticide and AChE. Mammalian studies spanning several decades have shown that anticholinesterase toxicity increases when CaE enzyme activity is inhibited. Although few studies are documented in fish, exposures to OP and carbamate pesticides have been found to reduce liver CaE activity in salmonids, with the chlorpyrifos acting as a more potent inhibitor of CaE activity than AChE activity. In another aquatic species (*Daphnia magna*), pharmacologic inhibition of CaE significantly enhanced the toxicity of chlorpyrifos, malathion, and carbofuran. Thus, although other biochemical targets may be involved in OPs and carbamates synergism, future mechanistic studies should give particular consideration to the role of CaEs in the pesticide interactions observed (Laetz *et al.*, 2009).

In an experimental study, acute toxicity of diazinon (an OP pesticide) was studied in *Anguilla anguilla* (European eel) fish. The median lethal concentration (LC_{50}) values of diazinon were found to be 0.16, 0.11, 0.09 and 0.08 mg/L at 24, 48, 72 and 96 hr exposure, respectively. The els were exposed to 0.056 mg/L of diazinon and the bioaccumulation and elimination of this insecticide in liver, muscle, gill and blood tissues were studied. Higher accumulation capacity of the gill was seen for the first hour of exposure. Diazinon elimination from the selected tissues was rapid; diazinon levels were not detected in any tissue after 24 hr in clean water. The half-lives of diazinon were calculated as 30.6, 32.2 and 38.3 hr for liver, muscle and gill, respectively (Sancho *et al.*, 1992). Histopathologic lesions in the liver of juvenile *Prochilodus lineatus* were assessed following 24 and 48 hr exposure to the OP pesticide Dipterex 500® (trichlorfon) at a concentration of 0.2 µl/L. Alterations in hepatic tissue were observed after pesticide exposure, including lateral migration of nuclei, variation in the diameter and density of nuclei, pyknosis and necrosis (Rodrigues *et al.*, 2001).

Dimethoate, a broad spectrum OP insecticide, is a potential toxic pollutant, adversely affecting the fauna of aquatic ecosystem. The fingerlings of common carp (*C. carpio*) were exposed to dimethoate. The LC_{50} values of dimethoate for 24, 48, 72 and 96 were found to be 1.84, 1.78, 1.68 and 1.61 mg/L, respectively. The test fishes showed the erratic swimming, increased surfacing, decreased rate of opercular movement, copious mucous secretion, reduced agility and inability to maintain normal posture and balance with increasing exposure time. Dissolved oxygen content of the troughs was measured at 24, 48, 72 and 96 hr to assess the impact of toxicant exposure on oxygen consumption. The oxygen consumption of exposed fishes showed significant decline at all concentrations. Therefore, dimethoate is highly toxic to fingerlings of common carp and severely affects their physiology and behaviour (Singh *et al.*, 2009).

L. rohita fishes were exposed to chlorpyrifos pesticide with the concentration of 0.1891 ppm for 2 days. Behavioural and morphological changes observed in the fishes might be due to formation of amino acids by degradation of protein with concentration of chlorpyrifos. Acute toxic effects of chlorpyrifos were also seen on haematological and biochemical parameters (Srinivasarao *et al.*, 2010). The static renewal evaluation of acute toxicity of an OP pesticide, dimethoate against freshwater fish *Heteropneustes fossilis* (catfish) was done. *H. fossilis* showed behavioural changes; there were increased opercular movement, sluggish, lethargic and abnormal swimming, loss of buoyancy and muscular tetany. The dimethoate-treated fishes also showed fading of their body colour. The LC_{50} values of dimethoate to *H. fossilis* at various exposure periods were 15.92 mg/L for 24 hr; 13.42 mg/L for 48 hr; 12.39 mg/L for 72 hr and 11.34 mg/L for 96 hr. The upper confidence limits were 16.59, 14.72, 13.54 and 12.86 mg/L for 24, 48, 72 and 96 hr and lower confidence limits were 15.32, 12.18, 11.23 and 9.79 mg/L, respectively. It seemed that the freshwater catfish, *H. fossilis* is more susceptible to dimethoate toxicity as the LC_{50} value for this is less than other reported fish species. The results indicated that dimethoate exposure to the fish caused toxic effects (Srivastava *et al.*, 2010).

Thiamethoxam (dimethoate) OP pesticide had significant impact on the growth and liver total protein of *O. niloticus* (Trewavas, a common freshwater exotic fish) (Bose *et al.*, 2011). Acute toxicity of an OP pesticide profenofos (O-4-bromo-2-chlorophenyl-O-ethyl S-propyl phosphorothioate) to freshwater fish, *C. punctatus* (Bloch), was studied in a static bioassay. Estimated 96-hr LC_{50} of profenofos was found to be 2.68 µg/L^{-1}. On the basis of the obtained LC_{50} values for 96 hr exposure intervals, profenofos can be rated as highly toxic to *C. punctatus*. Fish exposed to profenofos showed hyper excitability, discoloration, erratic swimming and secretion of excess amounts of mucus on the body and gills with eventual exhaustion and death (Pandey *et al.*, 2011). Toxic effects of monocrotophos OP pesticide were noticed in *C. carpio* fish. The oxygen consumption was decreased at lethal exposure (86.4 µg/L) over control on the 4th day, followed by 2nd, 3rd and 1st days. At 1/10th (8.64 µg/L) of lethal concentration, continuous decrease was seen on the 1st, 10th and 20th days but decrease was reduced on the 30th day. Similar variations were observed at 1/15th

(5.76 µg/L) of lethal exposure from day 1 to day 10. The decrease continued on 30th day which was very less amongst all. The behavioural responses were significant in lethal exposures due to monocrotophos intoxication. Fish showed hyper excitation, erratic swimming, jumping and lethargy due to low breathing frequency, followed by muscular rigidity and abundant mucous secretion over the gills (Maniyar *et al.*, 2011). The *C. punctatus* fish were exposed to sublethal concentration (0.019 ppm) of triazophos (an OP pesticide) for 24, 48, 72 and 96 hr to assess the alterations in the level of some biochemical parameters in blood plasma. Significant alterations in biochemical parameters were found to be dose dependent. Hypoglycemia and hypocholesteremia were observed in plasma of fish at the exposure periods. Increased activities of glutamate oxalacetate transaminase (GOT), glutamate pyruvate transaminase (GPT), acid and alkaline phosphatase of blood plasma indicated the hepatic tissue damage. Decrease in lactate dehydrogenase (LDH) content in plasma further indicated lower metabolic rate after 96 hr of exposure. Significant increase in triglycerides content was observed in fish exposed to 24 to 96 hr sublethal concentration of triazophos (Naveed *et al.*, 2011).

The LC_{50} value of malathion OP pesticide in freshwater catfish *H. fossilis* was found to be found to be 0.98 ppm for 96 hr. Histopathological changes in the liver, kidney and ovarian tissues of *H. fossilis* were observed after exposure of the sublethal concentration of 0.2 ppm (Deka and Mahanta, 2012). The effects of abate OP pesticide on the ovary of *H. fossilis* (catfish) was studied by observing the vitellogenesis of the fish. The LC_{50} of abate (0.64 ppm) after 24, 48, 72 and 96 hr of exposure to the fish was determined to be 14.3, 10.0, 6.61 and 5.13 ppm, respectively. In abate treated fish, the biochemical parameters, *viz.*, glycogen, protein and cholesterol were decreased. This pesticide potentiality altered the vitellogenesis of the fish (Kumari, 2012). Toxicity of kethrin (a botanical pesticide) and dichlorvos (an OP pesticide) was studied in *L. rohita*. The tests revealed LC_{50} values of 21.68, 16.71 ppm for kethrin and dichlorvos, respectively at 96 hr. The fish exhibited erratic swimming, copious mucus secretion, loss of equilibrium and hitting to the walls of test tank prior to mortality. Kethrin was less toxic to fish as compared to dichlorvos. The plant-based pesticides are biodegradable and are target specific than the highly persistent broad spectrum synthetic chemicals (Bhat *et al.*, 2012). The acute toxicity and effects of diazinon on some haematological parameters of kutum (*Rutilus frisii kutum*) fish were studied. The results showed that the LC_{50} value of diazinon was 0.4 mg L^{-1} at 96 hr and that the maximum allowable concentration value of this toxin was 0.04 mg L^{-1}. The long-term exposure to diazinon causes a decrease (P<0.05) in the RBC, haemoglobin (Hb), packed cell volume (PCV), mean corpuscular volume (MCV), mean corpuscular haemoglobin (MCH), mean corpuscular haemoglobin concentration (MCHC), WBC, lymphocyte, testosterone, iron (Fe), sodium (Na), lactate dehydrogenase (LDH) and cholinesterase. Diazinon also causes an increase (P<0.05) in prolymphocyte, aspartate aminotransferase (AST or GOT), cholesterol, alkaline phosphatase and adrenaline. The long-term exposure to low concentrations of diazinon causes changes in some haematological and biochemical parameters of *R. frisii* male brood stocks (Mohammad Nejad Shamoushaki *et al.*, 2012).

The OP pesticide dichlorvos (2, 2-dichlorovinyl dimethyl phosphate) is commonly used as agricultural insecticide. However, it is extremely toxic to non-target organisms such as fish and hampers the health of fish through impairment of metabolism, sometimes leading to death (Das, 2013). A significant increase in serum cholesterol (hypercholesterolemia) and serum glutamate pyruvate transaminase (SGPT) was observed in *C. punctatus* fish exposed to metasystox OP pesticide (four different concentrations at 24 to 96 hr). Hypercholesterolemia and elevated SGPT due to metasystox toxicity referred as biomarkers for disorders like myocardial infarction, heart attack, hepatitis biliary obstruction, alcoholic cirrhosis, cancer, toxic hepatitis, etc. in human beings (Mishra *et al.*, 2013). The mortality in *Colisa fasciatus* fish was recorded at 24, 48, 72 and 96 hr exposure of dimethoate. The LC values of dimethoate for 24, 48, 72 and 96 hr were found to be 22.15 mg/L, 21.99 mg/L, 21.74 mg/L and 21.65 mg/L, respectively. Behavioural alterations like decreased opercular movement, surfacing, abnormal swimming, increased mucus secretion and loss of balance were also observed in the exposed fishes (Singh, 2013). The 96 hr LC_{50} of the most commonly used OPs (*viz.*, profenofos and triazophos) was determined in native *L. rohita* fingerlings. These pesticides were applied to fingerlings that had been grown under optimised standard conditions under a maintained static bioassay system. Probit analysis was used for the determination of LC_{50} values, which were ascertained as 6.64 and 0.32 mg/L for triazophos and profenofos, respectively. The acute toxic stress was observed with fishes exhibiting behavioural intoxication, including suffocation, lying on the bottom, erratic swimming, lethargy and downward movements and gulping prior to the mortality (Ghazala *et al.*, 2014).

Toxicity of Carbamate Pesticides in Fish

Various Carbamate Pesticides and their Contamination

The most important carbamate pesticides are carbaryl and carbofuran. The carbamate pesticides are also cholinesterase (ChE) inhibitors with a similar mechanism of action as OP pesticides (Pandey and Pandey, 2013; Yue and Dai, 2011). The carbamate pesticides and their toxic metabolites are shown in Figures 63–65.

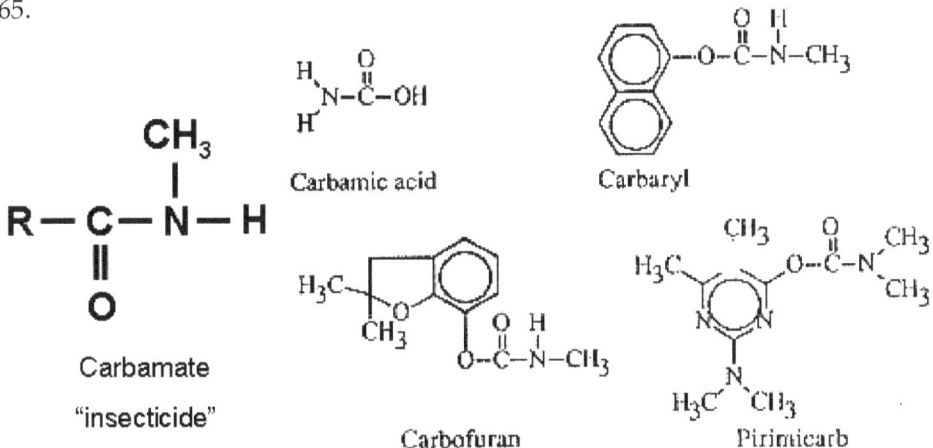

Figure 63–64: Carbamate Pesticides.

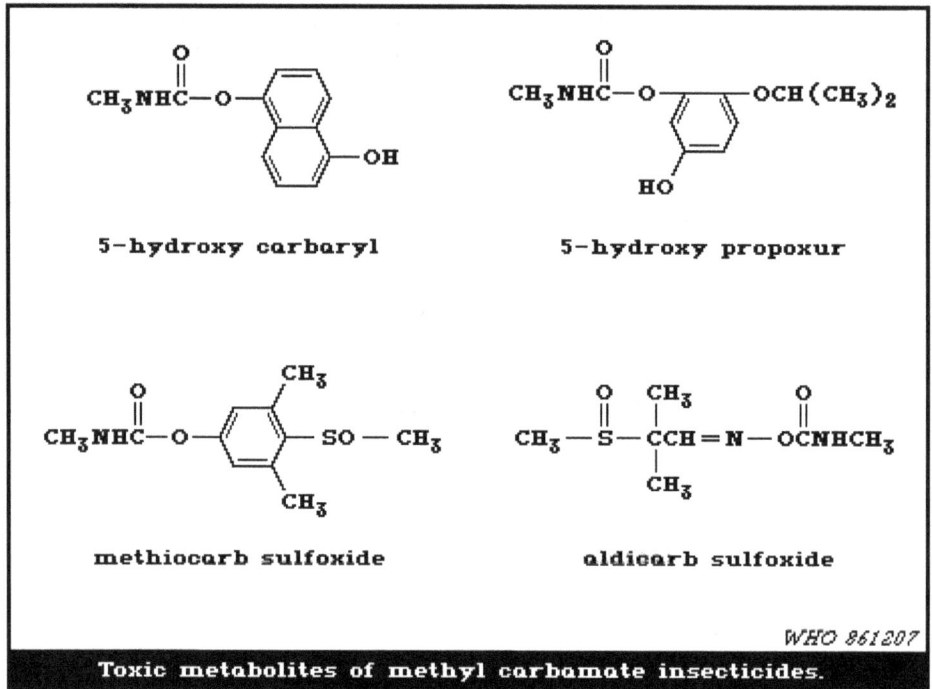

Figure 65: Carbamate Pesticides.

Contamination of carbamate occurs in the watersheds where the single-chemical assessments will systematically underestimate the actual risks to fish species, or aquatic organisms. Similar to OPs, the carbamate pesticides (insecticides) are water soluble and metabolized quickly. They generally have short persistence (half-lives of days to months), and their residues do not pose long-term problems for aquatic animals. Certain examples of carbamate pesticides are carbaryl and carbofuran. Many carbamates (*e.g.*, carbofuran) are extremely hazardous to fish and wildlife (Madhuri and Pandey, 2014).

The brain AChE inhibition was determined in juvenile coho salmon fish (*O. kisutch*) exposed to sublethal concentrations of the carbamates (*e.g.*, carbaryl and carbofuran), as well. The investigators observed addition and synergism of these pesticides, with a greater degree of synergism at higher exposure concentrations (Laetz *et al.*, 2009).

Toxicity of Carbamates in Fishes

Similar to OP pesticides, carbamates are also extremely toxic to fish and wildlife. They are water soluble and metabolized quickly. They generally have short persistence (half-lives of days to months), and their residues do not pose long-term problems for aquatic animals. Carbofuran (a carbamate pesticide) is extremely toxic to wildlife and fish (Laetz *et al.*, 2009; Madhuri and Pandey, 2014; Pandey and Pandey, 2013).

Toxicity of carbamate insecticide carbaryl (Seven®) and its metabolite, 1-naphthol, to four species of fish was studied. The calculated 96 hr LC_{50} values of carbaryl for *C. catla*, *Anabas testudineus*, *Mystus cavasius* and *M. vittatus* were 6.4, 5.5, 4.6 and 2.4 ppm, respectively, and that of 1-naphthol were 4.3, 3, 0.33 and 1.1 ppm in the respective fish species. The degradation product of the insecticide was found to be more toxic than the parent compound, to all the four species studied (Tilak *et al.*, 1981).

The effect of exposure to a sublethal concentration of a carbamate pesticide, sevin (1.05 mg/L) on biochemical parameters of blood, liver and muscle, and enzyme activities in liver, kidney, intestine, brain, gills and muscles of the freshwater teleost fish, *C. punctatus*, was studied after 15, 30 and 60 days. The changes produced were more marked after 30 and 60 days of exposure than after 15 days; the fishes were hyperglycaemic and hyperlactacidemic. The glycogen content of liver and muscles decreased, but lactic acid levels of the two tissues increased. In liver, muscles, brain and gills, the lactate dehydrogenase (LDH) activity was higher in pesticide-exposed fishes in comparison to control fishes, but the same enzyme activity was inhibited in kidney and intestine. Decreased pyruvate dehydrogenase (PDH) activity occurred in all six tissues. The succinate dehydrogenase (SDH) activity decreased in muscle at the three time periods, and after 30 and 60 days in liver and brain. However, in kidney and intestine, the latter enzyme activity was elevated. The results suggested that anaerobic metabolism was favoured and aerobic oxidation of pyruvate was impaired in the fishes exposed to sevin (Sastry and Siddiqui, 1982).

Carbaryl (n-naphthyl, 1-methyl carbamate; a carbamate pesticide) is being used widely to control a variety of pests of different crops. In India, it is being extensively used against the pests of cotton, tobacco, paddy, vegetable crops, etc. However, the NPs of carbaryl and 1-naphthol (the main degradation product of carbaryl, especially in the aquatic environment) have been reported to cause toxicity in the carp fish, *L. rohita* (Tilak *et al.*, 1981). A marked reduction in haemotological parameters of *T. guineensis* fish exposed to chloropyrifos was seen. This is an indication of the deleterious effects of chemical pollutant to the body fluid (Chindah *et al.*, 2004).

The acute toxicity of methiocarb in juvenile rainbow trout (*O. mykiss*, 3.25±0.79 g) was evaluated in glass aquaria under static conditions. Nominal concentrations of methiocarb in the toxicity test ranged from 1.25 to 7.50 mg L^{-1}. The concentrations of methiocarb that killed 50 per cent of the rainbow trout within 24 hr (24 hr LC_{50}), 48 hr LC_{50}, 72 hr LC_{50} and 96 hr LC_{50} were 5.43±0.19, 5.04±0.18, 4.95±0.19 and 4.82±0.21 mg L^{-1} (95 per cent confidence limits), respectively. The mortality of fish increased with increasing water temperature. Increasing alkalinity from 19 mg L^{-1} as $CaCO_3$ to 40, 60, or 90 mg L^{-1} as $CaCO_3$ significantly decreased mortality of the fish. Total hardness ranging from 50 mg L^{-1} as $CaCO_3$ to 147 mg L^{-1} as $CaCO_3$ did not affect mortality of the fish exposed to methiocarb. The fish exposed to methiocarb had histological changes like lamellar oedema, separation of epidermis from lamellae and lamellar fusion. The methiocarb exposed fish had necrosis between molecular and granular layer of cerebellum where the Purkinje cells present. Results indicated that alkalinity, temperature and fish size affected the methiocarb toxicity of rainbow trout (Altinok *et al.*, 2006).

The 96 hr LC_{50} of the most commonly used carbamates (*viz.*, carbofuran and carbaryl) was determined in native *L. rohita* fingerlings. These pesticides were applied to fingerlings that had been grown under optimised standard conditions under a maintained static bioassay system. Probit analysis was used for the determination of LC_{50} values, which were ascertained as 1.40 and 8.24 mg/L for carbofuran and carbaryl, respectively. Acute toxic stress was observed with subjects exhibiting behavioural intoxication including suffocation, lying on the bottom, erratic swimming, lethargy and downward movements and gulping prior to mortality (Ghazala *et al.*, 2014).

Detection of Toxic Nanoparticles of Organophosphates and Carbamates

Nanoparticles (NPs) of two most dangerous pesticides, the OP and carbamate are responsible for many hazards to fish species. Fish exposed to pesticides containing some of the most intensively used insecticides showed either concentration-additive or synergistic toxicity as well as unpredicted mortality. This implies that single-chemical assessments will systematically underestimate actual risks to fish species, or aquatic environment/organisms in the watersheds where the contamination/pollution of OP and carbamate NPs occurs (Pandey and Pandey, 2013).

Biosensors have been developed for the detection of pesticides using integrated enzymes, antibodies, cell and DNA-based biosensors. Enzymatic determination of pesticides is most often based on inhibition of the activity of selected enzymes such as ChE, acid phosphatase, ascorbate oxidase, acetolactate synthase and aldehyde dehydrogenase. Organophosphorus hydrolase (OPH) is an organophosphotriester hydrolyzing enzyme; the enzyme has broad substrate specificity and is able to hydrolyze a number of OP pesticides such as paraoxon, parathion, coumaphos, diazinon, dursban, etc. The OPH can be integrated with an amperometric transducer to monitor the oxidation or reduction current of the hydrolysis products. Many reports on integrated OPH enzyme for identification of different classes of pesticides (*e.g.*, carbamates and OPs) have been published. The ability of the nanoparticles based sensor to reliable measure concentrations of the OP pesticide carbofuran at nM concentrations was demonstrated by monitoring the inhibition of the hydrolysis of acetylthiocholine. Immobilization of AChE enzyme on multiwall carbon nanotubes and multiwall carbon nano-chitosan was proposed and thus a sensitive, fast and stable amperometric sensor for quantitative determination of OP insecticide was developed. An amperometric biosensor based on the adsorption of the AChE enzyme on screen printing electrodes (SPEs) and SPE coated with a Nafion layer were investigated. The sensor SPE was used to detect the inhibitory effects of OP and carbamate insecticides on AChE, and more particularly on chlorpyrifos ethyl oxon. A disposable carbon nanotube (CNT)-based biosensor was successfully developed and applied to the detection of OP pesticides and nerve agents. A new design of an enzyme biosensor based on AChE immobilized on the supported monomolecular layer composed of poly (amidoamine) of the fourth generation mixed with 1-hexadecanethiol was developed. The resulting enzymatic activity, measured amperometrically, was substantially depressed in the presence of the OP pesticide *e.g.,* dimethyl-2, 2-dichlorovinylphosphate (DDVP, dichlorvos), carbamate pesticides, *e.g.,* carbofuran and carbamate drug eserine. By using all these methods, the NP levels of organophosphate and carbamate pesticides can also be detected which may be helpful in knowing their toxic levels (Pandey and Pandey, 2013).

Furthermore, the ChE enzymes based biosensors have emerged as an ultrasensitive and selective technique for toxicity monitoring for environmental, agricultural, food or military applications. These devices are based on the inhibition of ChE by toxicants like pesticides. The main motivation for designing ChE biosensors for toxicity monitoring is to provide a reliable alternative to classical methods currently used in chromatographic methods. The electrochemical biosensors are presently among the most popular of the various types of biosensors. Carbon nanotubes (CNTs) are promising materials for sensing applications due to fascinating electronic and optoelectronic properties that are distinct from other carbonaceous materials and NPs of other types. Particularly, the properties of small dimensions, functional surfaces, good conductivity, excellent biocompatibility, modifiable sidewall, and high reactivity make CNTs have some overwhelming advantages in fabricating electrochemical sensors with high performances. Moreover, CNTs have an outstanding ability to mediate fast electron-transfer kinetics for a wide range of electroactive species, such as AChE. The CNT chemical functionalization can be used to attach almost any desired chemical species to them, which allows us to enhance the solubility and biocompatibility of the tubes. This has permitted the realization of composite electrodes comprising CNTs well-dispersed in an appropriate polymer matrix. Many enzymes used for the detection of pesticides are inhibited by the pesticide and the extent of inhibition is correlated to the concentration of the analyte. The AChE enzymes catalyze the hydrolysis of acetylcholine, an ester which is a neurotransmitter. The reaction catalyzed by AChE is: acetylcholine + $H_2O \rightarrow$ choline + acetate. The OP and carbamate pesticides are designed to inhibit AChE and this enzyme has been mostly used in enzymatic detection of these pesticides. The inhibition of AChE by OPs takes place as a result of the phosphorylation of the serine residue in the active site of the enzyme. The hydroxyl group on the serine residue acts as an electrophile which attacks the nucleophilic phosphorus. The phosphorylated enzyme is highly stable and the hydrolysis of acetylcholine is blocked. In some cases, depending on the chemical structure of the pesticide, the phosphorylation, and thus the inhibition, may be irreversible (Pandey and Pandey, 2013; Yue and Dai, 2011).

A highly sensitive, rhodamine B-covered gold nanoparticle (RB-AuNP)-based assay with dual readouts (colorimetric and fluorometric) has been recently described for detecting OP and carbamate pesticides in complex solutions. The detection mechanism is based on the fact that these pesticides can inhibit the activity of AChE, thus preventing the generation of thiocholine (which turns the RB-AuNP solutions blue and unquenches the fluorescence of RB simultaneously). This assay allows detection of pesticides in real samples such as agricultural products and river water (Liu *et al.*, 2012). Similar to the OP and carbamate NPs, the acute toxicity of colloidal silver nanoparticles (AgNPs) was tested in to determine the significance of the life stage of rainbow trout fish for nano-eco-toxicological study. The, colloidal AgNPs seemed to be 'very toxic' and 'toxic' to the eleutheroembryo-larva and juvenile stages, respectively, meaning that the release of nanosilver into the aquatic environment or its (silver NP) direct application as an antimicrobial agent in aquaculture should no longer be allowed. The ChEs, a class of serine hydrolases, are catalyzers of choline esters. Certain pesticides, *e.g.*, OPs and carbamates, are known to selectively inhibit ChE activity. Notwithstanding, an increase of brain ChE has been reported in *O. mykiss* and *Sparus auratus* fishes exposed to sub-lethal copper concentrations. Plus, when investigating the impact of *in vivo* exposure to metals (Cu, Cd, Hg, and Zn) on the AChE activity in fish, the results showed both a decrease and increase of AChEs during metal exposure, depending on the tissue sampled and the duration of the exposure (Johari *et al.*, 2013).

Chapter 8

Some Toxicants or Pollutants Causing Cancer in Fish

Fish Chemical Carcinogenesis

Cancer in Different Fish Species

'*Cancer*' is a frightful disease of human and other multicellular animals in which patients suffer pain, disfigurement and loss of many physiological processes. Cancer may be uncontrollable and incurable, and may occur at any time at any age in any part of the body. '*Carcinogenesis*' is a process which results in the appearance of lesions that meet the requirement of a malignant tumour (cancer). It is caused by a complex, poorly understood interplay of genetic and environmental factors. It represents the largest cause of mortality in the world as the millions of humans and animals die every year. Cancer is the abnormal growth of cells in our bodies that can lead to death. Cancer cells usually invade and destroy normal cells. These cells born of an imbalance in the body and by correcting imbalance, the cancer may go away. Billions of dollars have been spent on cancer research and yet we still don't understand exactly what cancer is (Madhuri and Pandey, 2014; Pandey *et al.*, 2013a).

A large number of agents cause genetic damage and neoplastic transformation of cells. Chemical carcinogens, radiant energy, and oncogenic viruses and some other microbes can cause various types of cancer. Radiant energy and some chemical carcinogens are the documented causes of cancer in humans, and evidence linking certain viruses to human cancer grows ever stronger. A number of carcinogen classifications deal with groups of substances, *e.g.*, aniline and homologs, chromates, dinitrotoluenes, As and inorganic As compounds, Cd compounds, Ni compounds, crystalline forms of silica, beryllium and its compounds, organophosphorus compounds and pesticides. There are also substances of variable or unclear chemical makeup that are considered carcinogens. Some of these substances include coal tar pitch volatiles, coke oven emissions, diesel exhaust and tobacco smoke, etc. (Kumar *et al.*, 2006; Pandey *et al.*, 2013a).

Like other animals and humans, fish can also suffer from various types of cancer. Many cancers in fish appear to be the result of exposure to different environmental chemicals/pollutants. High frequencies of liver and skin cancers in brown bullheads are associated with the high concentrations of PAHs and some metals in the environmental sediments. Data correlated with pollutant exposure and laboratory induction experiments reinforce the idea of chemical carcinogenesis in wild fish population. Liver neoplasms in sauger and walleye are associated with heavy loadings of extremely fine particulates which were produced when copper *'stamp sands'* were reprocessed. Now, one in four of the St. Lawrence whales are dying from cancer, mostly intestinal cancer. When scientists examined their bodies, the autopsies revealed high levels of PAHs. Epidemics of liver cancer were seen in 16 species of fish in 25 different polluted freshwater and saltwater locations. The same tumours were seen in the bottom-feeding fishes in industrialized and urbanized areas (Madhuri *et al.*, 2012c and 2012e; Pandey *et al.*, 2013a).

Therefore, pollution of rivers and streams with chemical contaminants has become one of the most critical environmental problems. Fish living in a polluted water reservoir use contaminated water to rinse their gills, resulting into deposition of PAHs in the fish body. Contamination of foodstuffs by heavy metals (*e.g.*, As, Cd, Cr, Ni and Pb) has posed a potential carcinogenic threat to humans. The As and Cd appear to be the most harmful to fish (Madhuri *et al.*, 2012c; Pandey *et al.*, 2013a). Thus, the most important chemicals causing severe toxicity to fish are heavy metals, *e.g.*, As, Cd, Pb, Ni, Cu, Mn, Cr, Fe, Hg, Zn, Sn, etc. These chemicals can also cause fish cancer. Besides these, DEN, high concentrations of PAHs and some other metals cause the cancer of liver, skin and other organs of fish (Madhuri *et al.*, 2012e and 2012f; Pandey *et al.*, 2013a).

Scientific and public interest in the occurrence of tumours (neoplasms) began slowly in the 1960s; and by 1986, systematic tumour surveys started in some parts of the world. Contaminant surveillance programmes have reported that the environment is receiving and accumulating an array of organic and inorganic contaminants. The presence of carcinogens, combined with fish tumour induction data and reports of tumours in wild fish populations, have provided strong circumstantial evidence that chemical carcinogenesis is occurring at some sites (Pandey *et al.*, 2013b). Liver has been found to be the most targeted organ for fish cancer. The cancers which have been reported in fish (particularly in zebrafish) are seminoma, HCC, HCA, adenoma of exocrine pancreas, intestinal adenocarcinoma, ultimobranchial neoplasm, thyroid neoplasm, spindle cell sarcoma and hemangioma (Madhuri *et al.*, 2012h).

A great attention has been given on fish over the past three decades to have as valuable fish models for research in carcinogenesis. Various fish species have been investigated as non-mammalian vertebrate models for carcinogen testing, as surrogates for understanding mechanisms of human cancer and its prevention, as feral species indicators of ecologic contamination, as indicators of potential human exposure to carcinogens in the water column or aquatic food chain, and for application as *in situ* field monitors of integrated carcinogenic hazard in groundwater near toxic waste sites. Compared to mammals, cancer induction in fish appears to be easy. The carcinogens can be exposed through aquarium water and liver cancer induction, in particular, provides best results with most of the carcinogens so far studied (Raisuddin and Lee, 2008).

In the 1960s to mid-1970s, the fish species used as models for carcinogenesis studies were primarily the zebrafish (*D. rerio*) and the guppy (*P. reticulata*). Species that have predominated in later years can be divided into two groups: the larger fish, *e.g.*, rainbow trout, and the small aquarium fish, including rivulus (*R. marmoratus*), guppy (*P. reticulata*), sheepshead minnow (*C. variegatus*), platyfish/swodtail (*Xiphophorus/Platypoecilus* sp.), *F. grandis, Gambusia affinis,* and medaka (*O. latipes*). The contaminations associated neoplasia, including aflatoxin-induced HCC in rainbow trout (*O. mykiss*) fish, have also led to the study of fish as alternative models in carcinogenesis and toxicity bioassay (Bunton, 1996; Pandey *et al.*, 2014). Teleost fish cell lines have been developed from a broad range of tissues such as ovary, fin, swim bladder, heart, spleen, liver, eye muscle, vertebrae, brain and skin. The 124 new fish cell lines from different fish species ranging from grouper to eel have been reported. Recently, about 283 cell lines have been established from finfish around the world (Lakra *et al.*, 2011; Pandey *et al.*, 2014).

Further, zebrafish is increasingly recognized as a promising animal model for cancer research, and liver is the main target organ for tumorigenesis, especially HCC after carcinogen treatment. Zebrafish is a responsive, cost-effective lower vertebrate model system to study the mechanism of carcinogenesis. Liver has been found to be the most commonly targeted organ in the cancer studies with zebrafish. The neoplasms which were noticed in zebrafish include seminoma, HCC, HCA, adenoma of exocrine pancreas, intestinal adenocarcinoma, ultimobranchial neoplasm, thyroid neoplasm, spindle cell sarcoma and hemangioma (Madhuri *et al.*, 2012h; Pandey *et al.*, 2013a and 2014).

Main Chemical Carcinogens

The main chemical carcinogens which have been reported (Kumar *et al.*, 2006; Madhuri and Pandey, 2009; Pandey *et al.*, 2013a) to cause carcinogenesis/cancer are shown in Table 3.

The chemicals, *e.g.*, DDT, kepone, methoxychlor and toxaphene (all are insecticides), atrazine (a herbicide), PCBs and DES, which possess oestrogenic activity can act as endocrine disruptors and may lead to the development of many cancers (Mukherjee *et al.*, 2006; Pandey *et al.*, 2013a). Many studies showed that several environmental factors, including air, water and industrial pollutants, environmental chemicals and radiations, etc. may cause various types of cancer. Increased infections, more use of pesticides, low consumption of fruits and vegetables, increased consumption of alcohol and red meat, more smoking, high industrial pollution, more exposure to sun, decreased physical activity and high occupational exposures may cause different types of cancer. The major portion of chemicals to which humans are

Table 3: Main Chemical Carcinogens

Sl.No.	Chemical Carcinogens		
1.	Direct actingcarcinogens	(a)	Alkylating agents- b-Propiolactone, dimethyl sulphate, diepoxybutane and anticancer drugs (cyclophosphamide, chlorambucil, nitrosoureas, etc.)
		(b)	Acylating agents- 1-acetyl-imidazole and dimethyl carbamyl chloride
2.	Indirect actingcarcinogens orprocarcinogens	(a)	Polycyclic and heterocyclic aromatic hydrocarbons- Benz(a)anthracene, B(a)P, dibenz(a,h)-anthracene, 3-methyl-cholanthrene and 7,12-dimethyl- Benz(a)anthracene
		(b)	Aromatic amines, amides and azo dyes- 2-naphthylamine (b-naphthylamine), 2-acetyl-aminoflucrene, dimethyl-aminoazobenzene (butter yellow) and benzidine
		(c)	Natural plants and microbial products- Aflatoxin B_1 (AFB_1), griseofulvin, cycasin, safrole and betel nuts
		(d)	Others- Nitrosamine, amides, vinyl chloride, nickel, chromium, insecticides, fungicides and polychlorinated biphenyls (PBs)

exposed is naturally occurring, that are carcinogenic at large doses. Almost every fruit and vegetable contains natural carcinogenic pesticides. Many natural chemicals are ingested as carcinogens from cooking food like roasted coffee. These include caffeic acid, chlorogenic acid, catechol, DDT, furfural and B(a)P, etc. A diet free of naturally occurring carcinogenic chemicals is impossible. Food additives (*e.g.*, allyl isothiocyanate, alcohol, butylated hydroxyanisole and saccharin), mycotoxins (*e.g.*, aflatoxin and hepatitis B virus) and synthetic contaminants (*e.g.*, PBs and tetyrachlorodibenzo-p-dioxin) are also mutagenic and carcinogenic (Pandey and Madhuri, 2010b; Pandey *et al.*, 2013a). Pesticides have become a part of environmental contaminants due to their widespread use in agriculture and disease control. Many pesticides are immunotoxic and found to suppress the cell-mediated immunity (Mukherjee *et al.*, 2006; Pandey *et al.*, 2013a).

Carcinogenesis in Fish by Certain Chemical Compounds

Although carcinogenesis in fish is caused by several carcinogens, however, the following chemicals having strong carcinogenic effects in fish are discussed here.

1. Diethylnitrosamine (DEN)

This is a nitrosamine or N-nitroso compound, and is one of the most potent carcinogens. Primary neoplasms with histological characteristics of HCC were observed in fish exposed to 125 ppm of DEN for 3 to 5 successive periods (Pandey, 2011a). At the 18th week, hemangiomas, cholangiomas, biliary cystadenomas, and glandular, trabecular and anaplastic HCC in *R. marmoratus* fish exposed to 95 or 200 mg per L DEN for 6 weeks were observed. It was then discovered that the groundwater contaminated with trichloroethylene (TCE) has carcinogenic properties beyond what had been shown with TCE alone, suggesting that the unidentified compounds in the mixture might have promoting properties alone or synergistically with TCE. The rainbow trout were fed with a diet containing indole-3-carbinol (2000 ppm), β-naphthoflavone (500 ppm) or aroclor 1254 (100 ppm) for 6 weeks before a single 24 hr exposure of an aqueous solution of 250 ppm DEN. After 42 weeks, DEN produced 80.2 per cent incidence of liver tumours. The tumour was inhibited by indole-3-carbinol but enhanced by β-naphthoflavone. The liver cytotoxic alterations of adult medaka fish were seen following short-term bath exposure (48 hr) to 500 mg per L DEN for 3 to 21 days. Progression of hepatic neoplasia was observed in adult medaka fish (3-6 months old) following aqueous exposure to DEN (50 ppm for 5 weeks). DEN-induced spongiosis hepatis (a hepatic lesion characterized by mutilocular cyst-like complexes) was noticed in the medaka. The induction of extrahepatic tumours by DEN in fish is uncommon with the exception of pancreatic neoplasms, as seen in rivulus. DEN exposure for 2 hr was sufficient to induce the hepatic neoplasms in *K. marmoratus*. Tumour after exposure to carcinogen for such a short duration was amazing, as in other fish models it takes longer time and sometime repeated exposure is also needed. In zebrafish, about 8 weeks of exposure to DEN is needed to induce tumours. Carcino-embryonic antigen in DEN-induced liver, gut and biliary neoplasms in *K. marmoratus* fish suggested that the response was similar to mammals. Hepatomas were seen in 6- and 9- month groups of guppies (*P. reticulata*) treated with multiple doses of 100 ppm- and 200 ppm- DEN. Neoplastic foci of mixed hepatocytes and cholangiocytes increased in livers of guppies from 2nd month, developing into hepatoblastomas, which occurred in almost 100 per cent of guppies by 12th month (Pandey *et al.*, 2013a).

2. N-Methyl-N'-Nitro-N-Nitrosoguanidine (MNNG)

It is a direct-acting carcinogen that induces neoplasms of the gastrointestinal tract (GIT) in rats, hamsters and dogs with oral administration. As a group, the N-nitroso carcinogens cause liver neoplasia in essentially all fish species, when exposure occurs in early life stages. Liver neoplasms are also reported, as well as fibrosarcoma with subcutaneous injection and multiple organ involvement with intraperitoneal injection. In addition to liver neoplasia, MNNG causes neoplasia in gill, pseudobranch, thyroid, gonad, pancreas, gas bladder, mesothelium, scale, skin, olfactory epithelium, connective tissue, skeletal muscle, notochord, blood vessels and pigment cells of medaka. In rainbow trout, only epithelial neoplasms occur in MNNG-treated fish, with liver, stomach, swim bladder and kidney targeted in trout given aqueous exposures as embryos or fry. Neoplasia occurs only in stomach in rainbow trout fed MNNG, beginning at 2 months of age. Channel catfishes given subchronic aqueous exposure to MNNG beginning at 6 months of age develop neoplasia in skin, fins, gills, thymus, bone, oropharynx and generalized lymphoid system. A high incidence of thyroid neoplasia has been reported in rivulus following immersion exposure of young fry to MNNG. The MNNG also causes osteosarcoma of ribs and vertebrae in coho salmon treated as embryos. Embryos and fry were both quite responsive to MNNG; however, juvenile zebrafish were remarkably refractory to MNNG-induced neoplasia. The main target organs in zebrafish treated as embryos with MNNG were the liver and testis, with HCA as the most prevalent hepatic neoplasm. A variety of mesenchymal neoplasms occurred in zebrafish following embryo exposure to MNNG, including chondroma, hemangioma, hemangiosarcoma, leiomyosarcoma and rhabdomyosarcoma. The testis and blood vessels were the primary target organs for MNNG following fry exposure, with seminoma, hemangioma, hemangiosarcoma, and various other epithelial and mesenchymal neoplasms (Bunton, 1996; Pandey *et al.*, 2013a, 2013b and 2014).

With exposure of either juvenile or adult medaka, neoplasms were primarily induced in gills, connective tissues, skin, and olfactory and reproductive systems. Within the skin, neoplasms were variable and included melanoma. Vascular tumours were the most common type of neoplasm in one study, but many other tissues were affected. Multiple tissue involvement was mentioned in another juvenile exposure study, but only data on gill neoplasms were given. Despite the preceding variations, the tissue distribution of neoplasms suggests direct action of MNNG on exposed tissues. The reported distribution of neoplasms in rainbow trout exposed to MNNG was different from medaka, with no apparent relationship to exposure age or route. With embryos, fry, or adults exposed aqueously or by injection or diet, cancers were seen in liver, stomach and kidney, with a low incidence in swim bladder; however, in two of these reports, histologic examination was restricted to those same tissues. Liver cancers were common in rainbow trout; whereas in medaka, they were absent, not reported, or occurred at rates no higher than controls. In a separate report of 9,802 exposed (various compounds) and control medaka, 40 cases of lymphoma were seen, and lack of a significant difference between control and treated groups indicated that cancer may be spontaneous in this species. However, tumour biology of lymphosarcoma in catfish is unknown, and an 8.8 per cent incidence of lymphosarcoma in *Poeciliopsis* strain *P. monacha* exposed to DEN was seen. Most cancers were spermatocytic seminomas containing spermatocytes and spermatids. Fewer seminomas were comprised strictly of spermatogonia, as previously reported in medaka. Spermatocytic seminoma occurs in male and female medaka, in case of both spontaneous and carcinogen-related gonadal neoplasia (Bunton, 1996; Pandey *et al.,* 2013a, 2013b and 2014).

3. Benzo(a)pyrene [B(a)P]

This is an indirect-acting carcinogen. It can cause hepatocarcinogenic in rainbow trout after long-term exposure through diet or intraperitoneal administration. The racemic($\pm$)-trans-B(a)P-7,8-dihydrodiol is a much more potent carcinogen in trout. Reconstitution studies with purified enzyme and liver microsomes from BNF-treated trout indicate that CYPIA is the predominant subfamily involved in B(a)P and B(a)P-7,8-dihydrodiol bioactivation to the ultimate carcinogen 7S-trans-7,8-dihydrobenzo[a]pyrene-7,8-diol-anti-9,10-epoxide. Topical administration of a PAH extract or B(a)P also induced epidermal papillomas in two species, and a solitary gill hemangioma was seen in a high-dose B(a)P exposure group. The B(a)P is reported to cause liver cancer in guppy and madaka (Bunton, 1996; Pandey *et al.,* 2013a, 2013b and 2014).

4. Dimethylbenz(a)athracene (DMBA)

It is also an indirect-acting carcinogen. Although pancreatic neoplasms occurred, the incidence in DMBA- and B(a)P- exposed medaka was not higher than controls, and just one pancreatic adenoma was reported in the PAH extract exposure studies. It was suggested that the pancreatic acinar carcinoma may represent a spontaneous rather than carcinogen-induced neoplasm in medaka, but incidence of pancreatic carcinoma with MNNG exposure suggests that it may be carcinogen-dependent. A number of extrahepatic neoplasms were induced by DMBA in medaka, but the types were not specified. However, neoplasms were reportedly absent in guppy exposed to DMBA by intramuscular, intraperitoneal, or topical routes (Bunton, 1996; Pandey *et al.,* 2013a and 2014).

5. N-Methyl-N-Nitrosourea (MNU)

It is an alkylating agent which methylates the DNA bases primarily at nucleophilic sites (N^7 and N^3 alkylpurines). This is an indirect-acting carcinogen. The primary mutagenic lesion of MNU exposure is believed to be O^6-methylguanine. The MNU induces several cancers in rodents, including mammary carcinomas and thyroid tumours in rats. It has also induced various tumours in *Xiphophorus* hybrids, including neuroblastomas, melanomas, fibrosarcomas, rhabdomyosarcomas at high incidence, and various carcinomas at a greatly reduced incidence (Bunton, 1996; Pandey *et al.,* 2013a and 2014).

6. Other Chemicals

Cancers have been reported in brown bullheads in association with PAH and other contaminants. Other similarly affected benthic species include black bullheads (*Ictalurus melas*) exposed to chlorinated wastewater effluent, white suckers (*C. commersoni*) and white croakers. Increased incidence of papillomas is most commonly seen in polluted sites, papillomas have been seen in brown bullhead populations from relatively clean sites (Bunton, 1996; Pandey *et al.*, 2013a and 2014).

Certain other chemicals described in Table 4 can also develop the fish carcinogenesis/cancer (Bunton, 1996; Pandey *et al.*, 2013a and 2014).

Fish Cancer due to Certain Environmental Toxicants

Different Environmental Toxicants Involved in Fish Cancer

'Environmental pollution' covers all types of contamination emitted to the ground, surface water and atmosphere that could be further absorbed by plants, fish and farm animals. *'Pollution'* is produced by different *'environmental toxicants'*, such as industrial wastes, diesel exhausts (mostly in the vicinity of roads) and pesticide residues in food products, etc. An example of the effect of environmental pollution on nutrition is contamination of fish by PAHs. Another example is contamination of foodstuffs by heavy metals. Although some metals are essential for human nutrition, others including As, Cd, Cr, Ni and Pb have also posed a potential carcinogenic threat to humans. The As and Cd are the most toxic to fish (Madhuri *et al.*, 2012c).

Contamination of freshwater with chemicals or pollutants has become matter of concern over the last few decades. Aquatic systems may extensively be contaminated by heavy metals released from domestic, industrial and other man-made activities. Heavy metal contamination may have devastating effects on ecological balance of the recipient environment and a diversity of aquatic organisms. Because of the pollutants, a huge mortality occurs in fish. Different types of toxicity, leading to various diseases and cancer caused by heavy metals, *e.g.*, As, Cd, Pb and Hg is most common in fish. In cooperation with the US EPA, the *'Agency for Toxic Substances and Disease Registry'* in Atlanta, Georgia (a part of US Department of Health and Human Services) reported that in a *'Priority List for 2001'* called the *'Top 20 Hazardous Substances'*, As, Pb and Hg are at the 1st, 2nd and 3rd position, respectively in the list; while Cd is at the 7th place. The As is a common element that occurs in air, water, soil and all living tissues. It ranks 20th in abundance in the earth's crust, 14th in seawater and 12th in the human body. It is a carcinogen and can cause foetal death and malformations in different species of mammals (Madhuri *et al.*, 2012e).

The most important chemical contaminants causing severe toxicity to fish are heavy metals, *e.g.*, As, Cd, Cr, Ni, Pb, Cu, Fe, Mn, Hg, Zn and Sn, etc. Similarly, the high concentrations of PAHs and some metals in environmental sediments cause liver and skin cancers in brown bullhead fishes. Thus, the chemical contaminants are responsible for induction of carcinogenesis in fishes. The fishes living in a polluted water reservoir (in the vicinity of an oil refinery) use the contaminated water to rinse their gills, resulting into the deposition of PAHs in the fish body. The contamination

Table 4: Different Fish Cancers Caused by Certain Chemicals

Chemical Carcinogen	Fish Species	Type of Cancer
2-AAF	Lebistes reticulates (Guppy)	Hepatic cancer
Aflatoxin B_1 (AF B_1)	Poecilia reticulata (Guppy)	Hepatic cancer
	Oryzias latipes (Madaka)	Hepatocellular carcinoma (HCC), hepatic adenoma (HA)
	Oncorhynchus mykiss (Rainbow trout)	HCC
Butylated hydroxyanisole (BHA)	Rivulus ocellatus (marmoratus)	HCC
Dichlorodiphenyl-trichloroethane (DDT)	Oncorhynchus mykiss	HCC
Methylazoxymethanol acetate (MAMA)	Gambusia affinis	HCC, CCC
	Oryzias latipes	HCC, medullo-epithelioma
	Poecilia reticulata	Pancreatic acinar cell carcinoma (ACC), adenocarcinoma (AC)
N-nitrosodimethylamine (DMN)	Danio rerio (Zebrafish)	HCC, CCC, oesophageal cancer
	Poecilia reticulata	HCC, CCC
N-nitrosomorpholine(NM)	Poecilia reticulata	HCC, cholangiocellular carcinoma (CCC), oesophageal cancer
	Danio rerio (Zebrafish)	HCC, CCC, oesophageal cancer
Orthoaminoazotoluene(o-AAT)	Poecilia reticulata	HCC, CCC
	Oryzias latipes	HCC, HA

of foodstuffs is also by the heavy metals. Although some heavy metals are essential for human nutrition, others including As, Cd, Cr, Ni and Pb, have been found to pose a potential carcinogenic threat to humans as well as fish, and Cd appear to be the most harmful. In populations near urban centres where concentrations of PAHs, PCBs, DDT and other compounds were elevated in tissue (stomach and liver) and sediment, there was an increased chemical-associated risk of developing cancers and associated lesions (Pandey *et al.*, 2013b).

Distribution of heavy metals varies between fish species, depending on age, development status and other physiological factors. Fish accumulate substantial concentrations of Hg in their tissues and thus can represent a major dietary source of this element for humans. Fish are the single largest sources of As and Hg for humans. The primary sources of Hg contamination in humans are through eating fish. Metal contaminations in food, especially in marine products, have been broadly investigated (Emami Khansari *et al.*, 2005). Among animal species, fishes are the inhabitants that cannot escape from the detrimental effects of these pollutants. The studies carried out on various fishes have shown that heavy metals may alter the physiological activities and biochemical parameters both in tissues and in blood (Vinodhini and Narayanan, 2008). Liver, kidney and gill of fish are the main organs for pollution. Liver and gills as main organs for metabolism and respiration are target organs for contaminants accumulation, and after the exposure of pollutants, the damage to organs and tissues occurs (Madhuri *et al.*, 2012e; Montaser *et al.*, 2010).

The heavy metals due to their toxicity and accumulative behaviour can make different changes in aquatic environment like species diversity. Heavy metals can enter into water via drainage, atmosphere, soil erosion and all human activities by many ways. With increasing heavy metals in the environment, these elements enter the biogeochemical cycle. They can enter from contaminated water into fish body by different routes and accumulate in various organs. Some heavy metals, *e.g.*, Fe, Cu, Zn and Mn are essential for biological systems like enzymatic activities; whereas, other heavy metals like Pb, Cd and Hg have no known important role in living organs and are toxic even in trace amounts. Essential metals must be taken up from water, food or sediment by fish for its normal metabolism; however, these metals can also have toxic effects at high concentration (Dobaradaran *et al.*, 2010). Metal pollution may damage marine organisms at the cellular level and possibly affect ecological balance. Exposure and ingestion of polluted marine organisms as sea foods can cause health problems, including neurological and reproductive problems in people and animals. Chemicals of industrial effluents and products of ships and boats, which find their way into different water systems, can cause toxic effects in aquatic organisms. Petroleum products are one of the most relevant pollutants to aquatic ecotoxicology. Exposure to crude oil and derivatives can induce toxic symptoms in experimental animals. Petroleum hydrocarbons can act as a mediator in free radical generation in fish. The coastal pollution occurs with several pollutants, including heavy metals (Montaser *et al.*, 2010).

Now, it is necessary to know how to examine the fish for cancers, to recognize gross signs of tumours and know how to prepare samples for histopathology. Therefore, the field biologist should develop skill in recognizing the presence of commonly occurring cancers in fish and in preparing samples for further diagnosis. Zebrafish are used to study the development, toxicology and toxicopathology. Zebrafish is known for its rapid development in the laboratory. Expansion of synthetic chemical producing industries during the 1940's coincided with a number of pollution-associated fish neoplasia epizootics, with PAHs as significant components of contaminated sediments in several cases. Epizootics of primarily liver and skin neoplasia in benthic species near coastal urban or industrial areas indicated the sensitivity of fish species to known mammalian carcinogens. Potential for application of research findings to both human and environmental health issues make fish species attractive and valuable alternative models in the carcinogenesis and toxicity research (Pandey *et al.*, 2012c).

Important Toxicants Causing Fish Cancer

MNNG (a direct-acting toxicant/pollutant) induces many cancers, as stated earlier (Pandey *et al.*, 2013a). Carcinogenic effects of certain N-nitroso carcinogens have been studied in some fish species. A high incidence of thyroid neoplasia is reported in rivulus, following immersion exposure of young fry to MNNG. The MNNG also causes osteosarcoma of ribs and vertebrae in coho salmon treated as embryos. Benign or malignant thyroid neoplasia is the most common histologic type of epithelial neoplasia reported in medaka following fry treatment with MNNG; and seminoma is second most frequent histologic type of epithelial neoplasia seen in medaka given fry bath treatment with MNNG. The MNNG and other N-nitroso carcinogens induce pigment cell neoplasia in medaka, nibe croaker and *Xiphophorus* (Bunton, 1996; Pandey *et al.*, 2012c).

DEN, also called N-nitrosodiethylamine (DENA) is another carcinogenic pollutant/toxicant. It is one of the most potent carcinogens. Much of its carcinogenic effect has been discussed earlier (Pandey *et al.*, 2013a). In the 1970s, there was an increased frequency of liver cancer found in Norwegian farm animals, which were fed on herring meal, which was preserved using sodium nitrite. The sodium nitrite had reacted with dimethylamine in the fish and produced DEN. Nitrosamines can cause cancers in a wide variety of animal species, a feature that suggests that they may also be carcinogenic in humans. DEN has been extensively used in medaka and rainbow trout, and has proved to be a potent inducer of hepatic neoplasms (Bunton, 1996; Pandey *et al.*, 2012c).

Another potent carcinogen is MAMA, which acts both directly and indirectly. It causes liver, kidney and colon cancers in rodents and nonhuman primates when administered by different routes. B(a)P is an indirect-acting carcinogen, as described earlier (Pandey *et al.*, 2013a). B(a)P causes hepatocarcinogenesis in rainbow trout but its long-term exposure through the diet or intraperitoneally is needed (Bunton, 1996; Pandey *et al.*, 2012c and 2014). DMBA is another indirect-acting carcinogen, discussed earlier (Pandey *et al.*, 2013a). This pollutant/contaminant is much more potent hepatocarcinogen in trout than B(a)P and produces tumours in kidney, swim bladder and stomach as well. DMBA has been reported to cause HCC, hepatic adenoma, renal acinar cell carcinoma, neurilemmoma and fibrosarcoma in guppy. It has also caused HCC and lymphosarcoma in *Poeciliopsis* fish species. In *C. variegatus* also,

DMBA caused HCC and hepatic adenoma (Bunton, 1996; Pandey *et al.*, 2012c and 2014). MNU is an indirect-acting carcinogen as described earlier (Bunton, 1996; Pandey *et al.*, 2012c and 2013a). The papillary thyroid tumour induction by MNU has also been reported (Lee *et al.*, 2000).

Certain Chemical Toxicants Causing Hepatic and Skin Cancers in Fish

Frequent hepatic (liver) cancer and related lesions have been observed in mummichog (*F. heteroclitus*) fish from a PAH-contaminated site in the Elizabeth river along the mid-Atlantic coast. PAH sediment concentrations were associated with active and abandoned wood creosote treatment facilities near the site, which experienced several significant creosote spills in the 1960s. Using hepatoproliferative lesions in rodents as a guide, the lesions included eosinophilic, basophilic and clear cell foci of cellular alteration, HCA and HCC, and proliferative biliary lesions, which were not seen in populations from less contaminated sites. Epizootic hepatic neoplasia was also reported from the North Atlantic coast in winter flounder (*P. americanus*), a valuable, benthic (bottom dwelling) fish. The contaminated sites were associated with outfall from sewage treatment plants and dredge spoils. The dredge sediments contained a complex mixture of PAHs, PCBs, metals and other compounds. The hepatic lesions included basophilic foci of cellular alteration, HCA, HCC, cholangiocarcinoma and proliferative hydropically vacuolated ductular epithelial cells. The epizootic of hepatic neoplasia in another benthic fish, the English sole (*Parophrys vetulus*), on the opposite coast of the United States, have been recorded. In Puget Sound of Washington State, the cancers and associated lesions were seen only in fish from estuaries and embayments proximal to urban centers or pollution point sources. The sediments contained a wide range of chemicals, which included B(a)P and benzo(a)anthracene, heavy metals, PCBs and other chlorinated hydrocarbons. The lesions in juveniles included degenerative megalocytic hepatosis frequently accompanied by eosinophilic, basophilic and clear cell foci of cellular alteration. Benign and malignant hepatocellular and biliary cancers were seen in adult fish of at least two years of age. The English sole injected intramuscularly with B(a)P or a PAH-enriched fraction of sediment extract from Eagle Harbor also resulted in megalocytic hepatosis and foci of cellular alteration, providing a link between exposure and development of these lesions (Buntun, 1996; Pandey *et al.*, 2013b).

Various toxicants/pollutants have been reported to produce experimental liver cancer in fish. Primary neoplasms with histological characteristics of HCC were noted in fish exposed to 125 ppm of DEN for 3 to 5 successive periods. In another study, hemangiomas, cholangiomas, biliary cystadenomas, and glandular, trabecular and anaplastic HCC were observed at the 18th week in rivulus fish exposed to 95 or 200 mg per L DEN for six weeks (Bunton, 1996). The HCC was induced by DEN in medaka fish as early as five weeks post-exposure. The rainbow trout were fed with a diet containing indole-3-carbinol (2000 ppm), β-naphthoflavone (500 ppm) or aroclor 1254 (100 ppm) for six weeks before a single 24 hr exposure of an aqueous solution of 250 ppm DEN. After 42 weeks, DEN produced 80.2 per cent incidence of liver tumours. Liver cytotoxic alterations of adult medaka (*O. latipes*) fish were seen following short-term bath exposure (48 hr) to 500 mg/L DEN for 3 to 21 days (Bunton, 1996; Pandey, 2011a; Pandey *et al.,* 2013b). Progression of hepatic neoplasia was observed in adult medaka fish (3-6 months old) following aqueous exposure to DEN (50 ppm for 5 weeks) (Okihiro and Hinton, 1999). The DEN-induced spongiosis hepatis (a hepatic lesion characterized by mutilocular cyst-like complexes) was noticed in the Japanese medaka, *O. latipes* (a small aquarium fish) (Norton and Gardner, 2005). Induction of extrahepatic tumours by DEN in fish is uncommon with exception of pancreatic neoplasms, as seen in rivulus fish. DEN exposure for 2 hr was sufficient to induce hepatic cancer in *K. marmoratus*. This fish was used to study the oncogenes in biliary and hepatic cancers, and necrotic and regenerative phases of DEN-toxicity. Hepatomas were seen in DEN-exposed *K. marmoratus* and *P. reticulata* (Bunton, 1996; Pandey, 2011a; Pandey *et al.,* 2012c and 2013a).

Mechanisms of hepatocarcinogenesis in fish have also been studied. As the preneoplastic hepatic lesions in fish are in many ways similar to those noticed in rodent hepatocarcinogenesis, the terminology is the same. However, the relative significance of each lesion may vary in fish cancers. A number of nonspecific changes preceding the development of preneoplastic and neoplastic lesions are important in the pathogenesis of hepatic cancer in fish. In medaka, sheephead minnow and rivulus fishes exposed to DEN by the aqueous route, there is an early cytotoxic phase with necrosis and dropout of hepatocytes resulting in cystic degeneration. Vacuolar degeneration, fatty changes and hyalinized eosinophilic cytoplasmic inclusions may also develop and persist. The eosinophilic cytoplasmic inclusions described in DEN-exposed medaka, zebra danio, rivulus and MAMA-exposed gambusia most likely represent apoptotic bodies. The severity of necrosis appears to increase with chemical concentration but also occurs at lower exposure levels. The cytotoxic phase is followed by proliferation of epithelial and interstitial cells (Bunton, 1996).

The foci (basophilic, eosinophilic and clear cell types) of hepatocellular alteration morphologically similar to those described in rodent hepatocarcinogenesis have been reported in several fish species. All these three types of foci have been identified in hepatic neoplasia epizootics (Myers *et al.,* 1994). The basophilic foci, consisting of small RNA-rich cells as classically described in rodents, are considered to progress to HCC in rainbow trout, medaka, guppy and sheephead minnow. The eosinophilic foci, consisting of cells containing abundant smooth endoplasmic reticulum in rodents, have also been reported in several fish species. However, the differences in the normal tinctorial properties of male and female hepatocytes related to vitellogenesis can make differentiation of basophilic from eosinophilic foci difficult in medaka. The clear cell foci are most variable of foci of cellular alteration in fish species. The foci of cells containing abundant glycogen have been reported in rivulus, mummichog and sheephead minnow but less commonly in medaka and rainbow trout. The foci of vacuolated cells have been described as the most likely contained lipid in medaka, and a mixture of lipid and glycogen in rivulus. Proliferation of bile ductular epithelial cells and cholangiofibrosis are common findings, and compatible non-neoplastic counterparts of the commonly seen benign and malignant biliary neoplasms. Oval cells, a proliferative population of ductular epithelial cells common in rodent hepatocarcinogenesis are seen in the sheephead minnow and medaka exposed to

DEN, and these have been postulated as the cell type responsible for regeneration after cytotoxic dosages of aflatoxin B_1 in rainbow trout (Bunton, 1996). A distinct lesion of hydropic vacuolization of proliferative preductular cells seen in winter flounder and other benthic species is considered to be a biomarker of environmental chemical contamination. Correlation of this lesion with rodent oval cell hyperplasia has been suggested (Moore and Stegeman, 1994).

Proliferation of hepatic interstitial cells in response to carcinogenic exposure is an important component of fish carcinogenesis, perhaps more so than in rodents. Two lesions are of particular importance. *'Spongiosis hepatis'*, a multilocular pseudocystic lesion induced by N-nitroso compounds in rats, commonly occurs in fish hepatocarcinogenesis bioassays with a variety of compounds. It appears most commonly in the small aquarium species but is also seen rarely in benthic species from hepatic cancer epizootics (Myers *et al.*, 1994). The lesions first appear after the cytotoxic phase, with subdivision of cystic spaces by perisinusoidal cells, and can become extremely enlarged or aggressively multifocal. In both rats and fishes, the spongiosis hepatis may be associated with foci of cellular alteration or hepatocellular cancers but in fishes, the association appears to be random. Another interstitial cell lesion, *'spindle cell proliferation'* consists of an unspecified population of spindle-shaped cells which replaces portions of the hepatic parenchyma. Although spindloid biliary epithelial cells have been found to comprise portions of some lesions by keratin immunocytochemistry, a substantial population remains uncharacterized and may include fibroblasts, endothelial cells and/or perisinusoidal cells. Fibroplasia with deposition of collagen is uncommon in fish studies but a mesenchymal reaction, including cirrhosis has been noticed in medaka exposed to MAMA and other compounds. In medaka, the perisinusoidal cells have been shown to express actin in trabecular and schirrous HCC induced by DEN or MAMA. This change, also seen in mammals, is interpreted to be a phenotypic reaction to injury. However, perisinusoidal cells, spongiosis hepatis lesions and hyperplastic perisinusoidal cell aggregates in rats are also desmin positive, whereas comparable lesions in fish have been found negative, and the decreased desmin antibody sensitivity in fish tissues may have affected the results. The studies of DEN-induced toxic lesions and hemangiopericytoma have identified proliferative cells with features shared by both perisinusoidal cells and endothelial cells, raising the possibility of a common stem cell between these cell types. The most common type of interstitial cell neoplasm reported to occur in several species is hemangiopericytoma (Bunton, 1996; Pandey *et al.*, 2013b).

Epizootics of skin cancer/neoplasia have also been associated with chemical contamination. In 1941, Lucke and Schlumberger reported an epizootic of orocutaneous papillomas in brown bullhead (*Ictalurus nebulosus*) in the Delaware and Schuylkill rivers near Philadelphia. Since then, liver and skin cancers have been reported in brown bullheads in association with PAHs and other contaminants from the Great Lakes to New York State and Ontario (Canada). Other similarly affected benthic species include black bullheads (*Ictalurus melas*) exposed to chlorinated wastewater effluent, white suckers and white croakers (Bunton, 1996). Although increased incidence of papillomas is most commonly seen in polluted sites, the papillomas have been seen in brown bullheads from relatively clean sites (Poulet *et al.,* 1994). In addition, lesions on white suckers observed in the laboratory regressed under uncrowded conditions (Premdas and Metcalfe, 1994).

Papillomas have been induced in brown bullheads and mice by epidermal applications of heavily polluted sediment extract containing PAHs (Bunton, 1996). The epidermal neoplasms in brown bullheads appear to progress through the classic stages of skin carcinogenesis, including hyperplasia, papilloma and carcinoma (Poulet *et al.,* 1994), although carcinoma is less commonly seen. The endophytic hyperplasia appeared to progress to large fungating exophytic papillomas. Focal invasion through the basal lamina marked the transition from papilloma to carcinoma, similar to other report (Bunton, 1996; Pandey *et al.,* 2013c; Poulet *et al.,* 1994).

Gross Appearance of Various Fish Cancers

Fish have considerable evolutionary commonality with mammals, including man. Most of the diseases, including cancer occurring in the mammals, are also developed in fish. Cancer can produce many pathological signs (lesions) and symptoms in the fish. The pathological lesions of cancer can be observed macroscopically or grossly (by naked eyes) and under microscope. Here are some gross appearances (images/photos) of various fish cancers caused by different substances, such as drugs, chemicals, microorganisms, toxicants, poisons, pollutants or contaminants, etc. These images (Figures 66–80) have been cited from various internet websites (Fish Cancer Website, 2015; Pandey *et al.,* 2014) that are duly acknowledged.

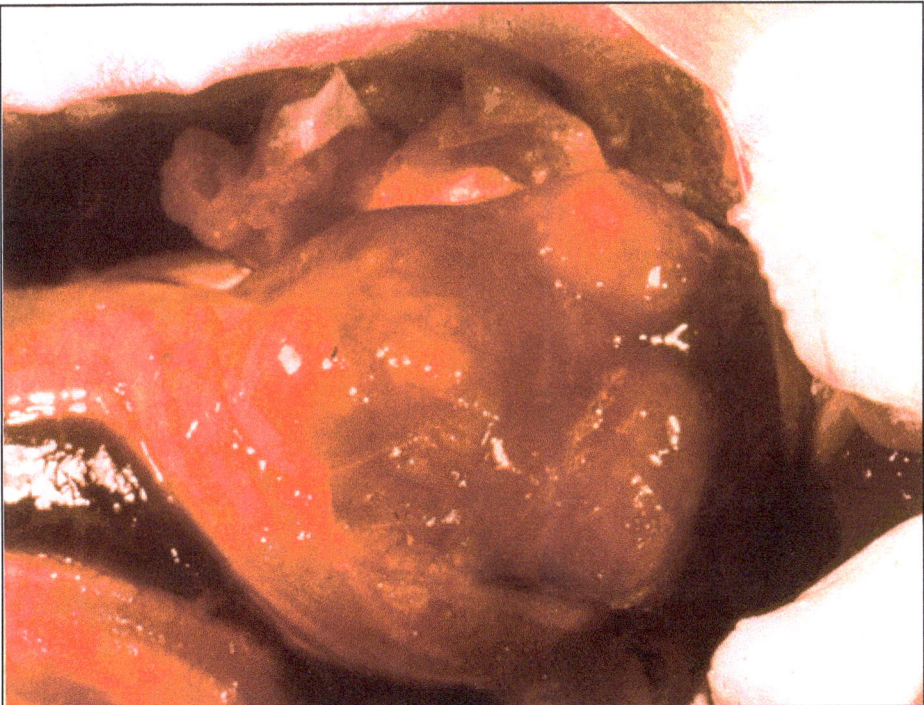

Figure 66: Cholangiolar and Hepatocellular Tumours in White Sucker.

Figure 67: Skin Cancer in Fish.

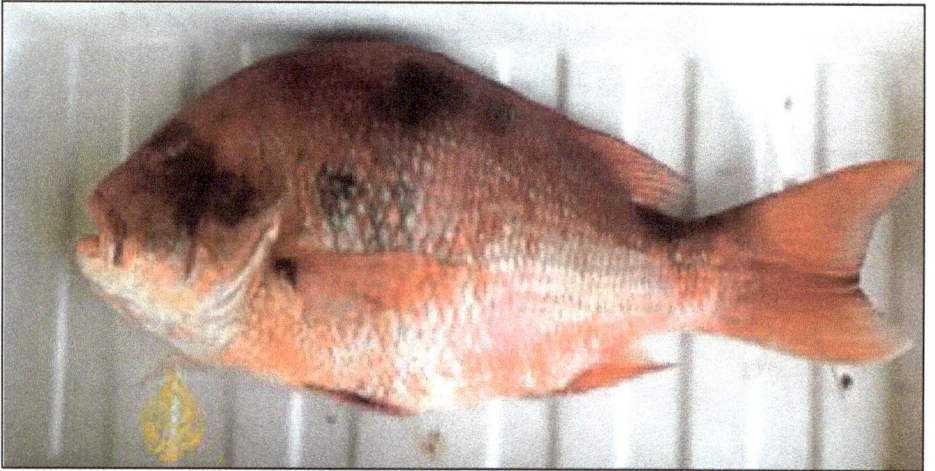

Figure 68: Skin Cancer in Walleye Fish.

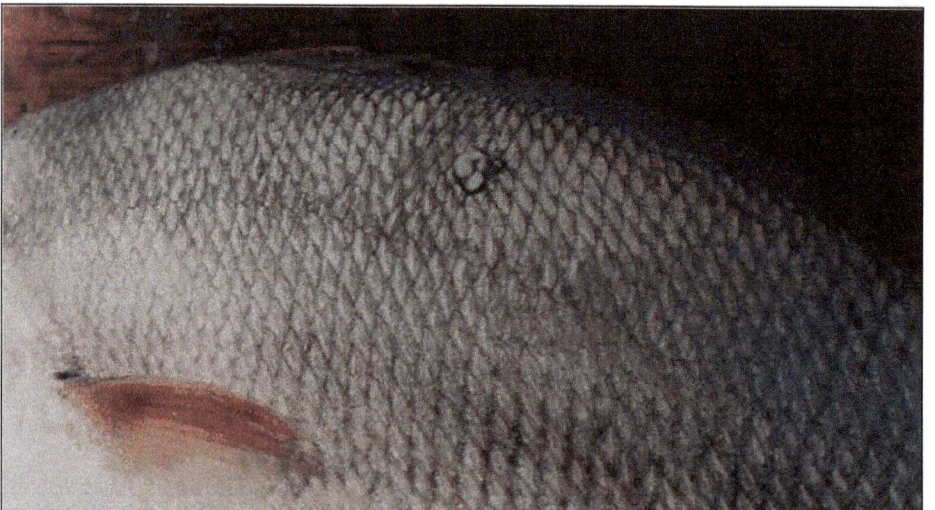

Figure 69: Dermal Cancer in Freshwater Drum.

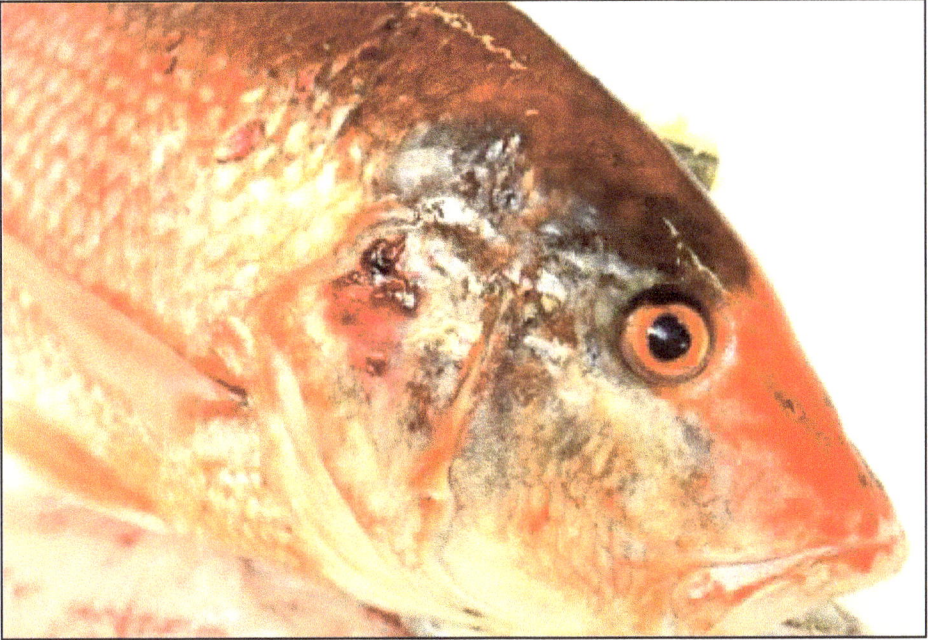

Figure 70: Cancer on Head Region of Fish.

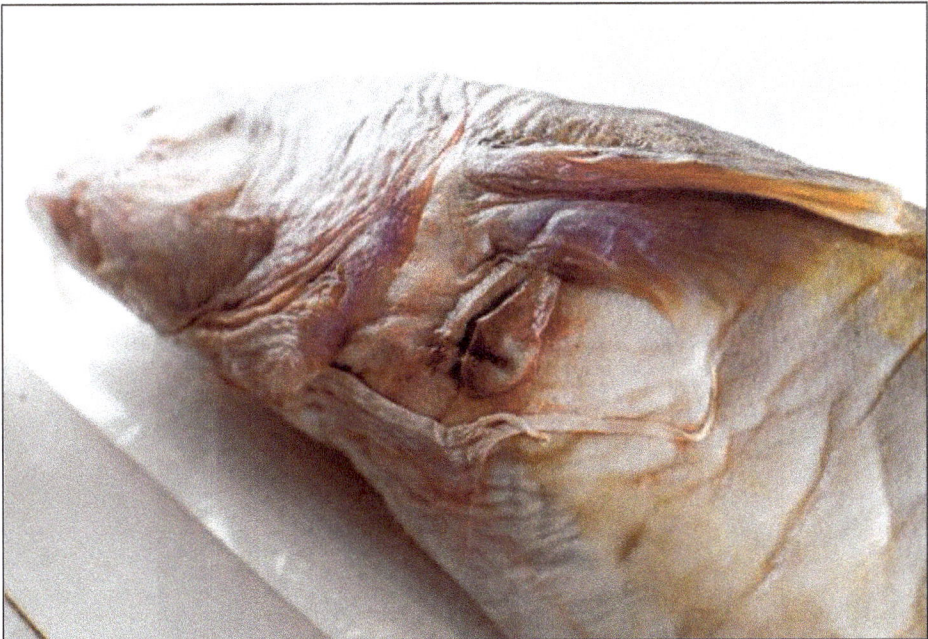

Figure 71: Cancer in Gills of Fish.

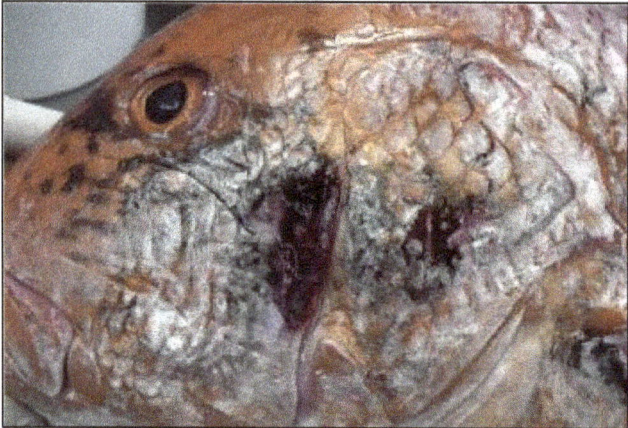

Figure 72: Cancer over Gill Region of Fish.

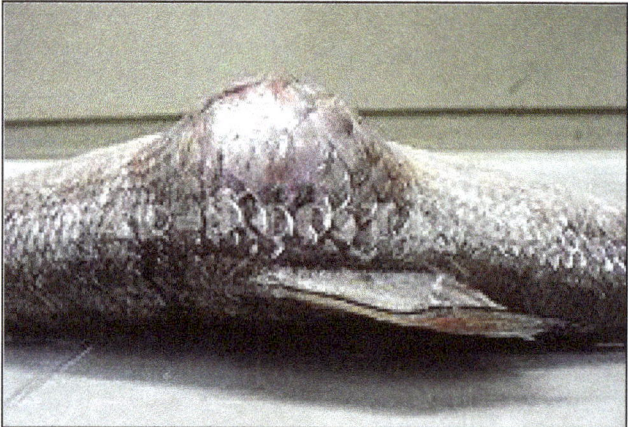

Figure 73: Tumour/Cancer in Fish.

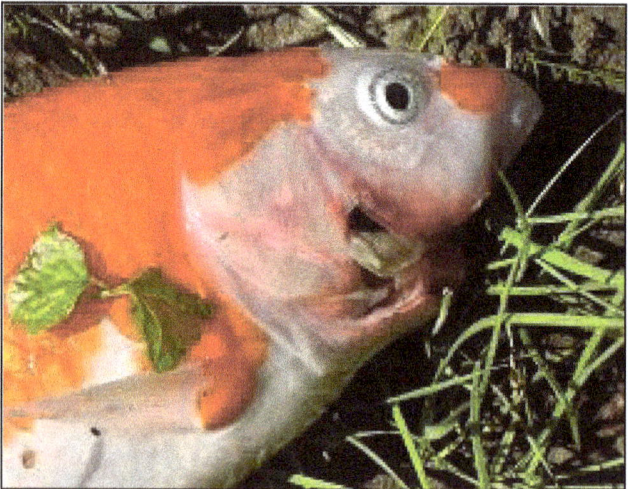

Figure 74: Moth Cancer in Fish.

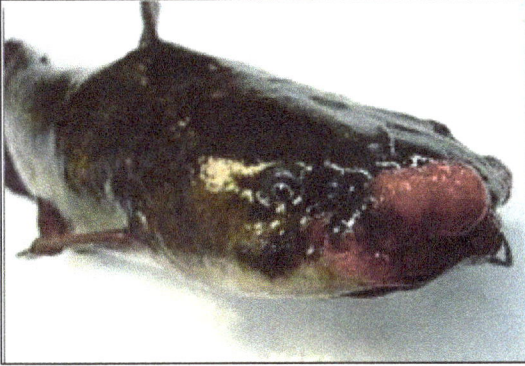

Figure 75: Cancerous Lesions in Mouth of Fish.

Figure 76: Cell Tumour in Fish.

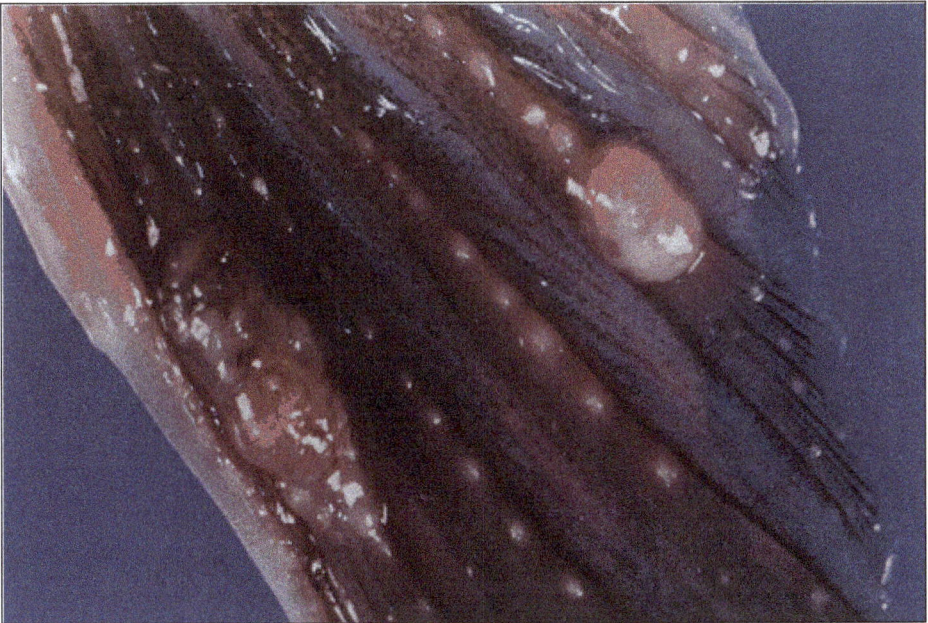

Figure 77: Papilloma on Fin of White Sucker.

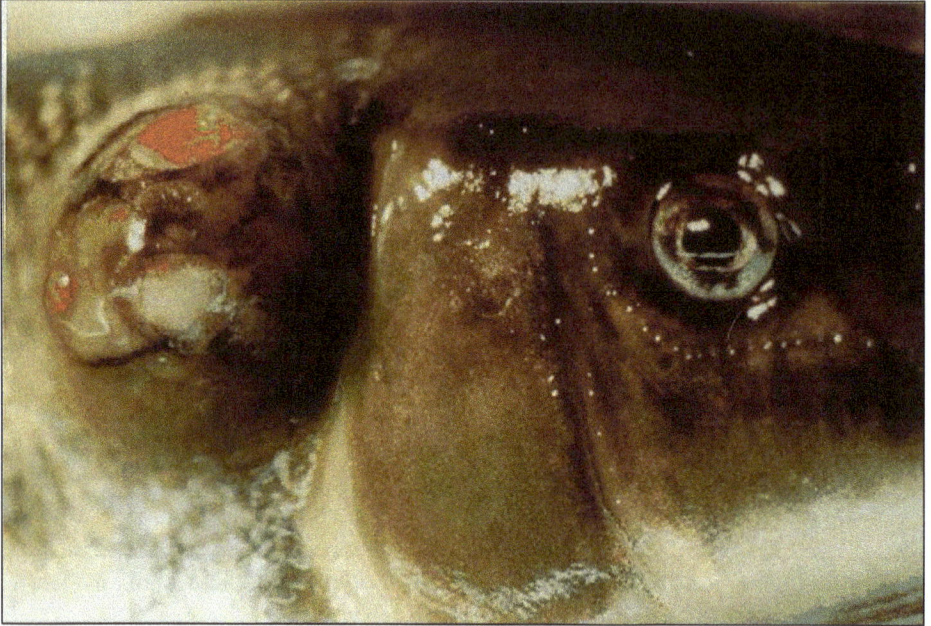

Figure 78: Fibroma in White Sucker.

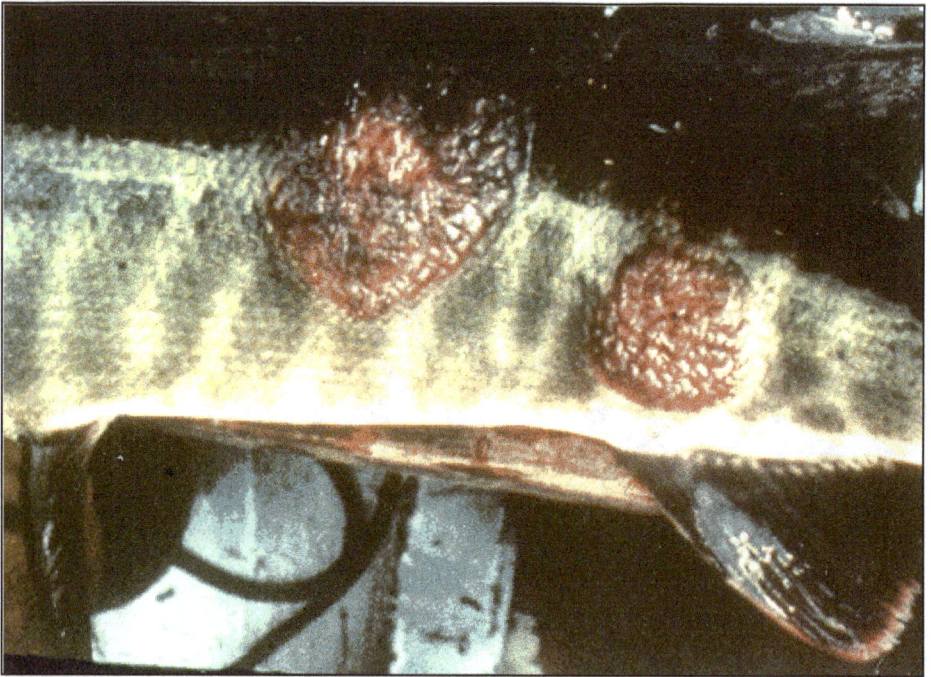

Figure 79: Lymphosarcoma in Muskellunge.

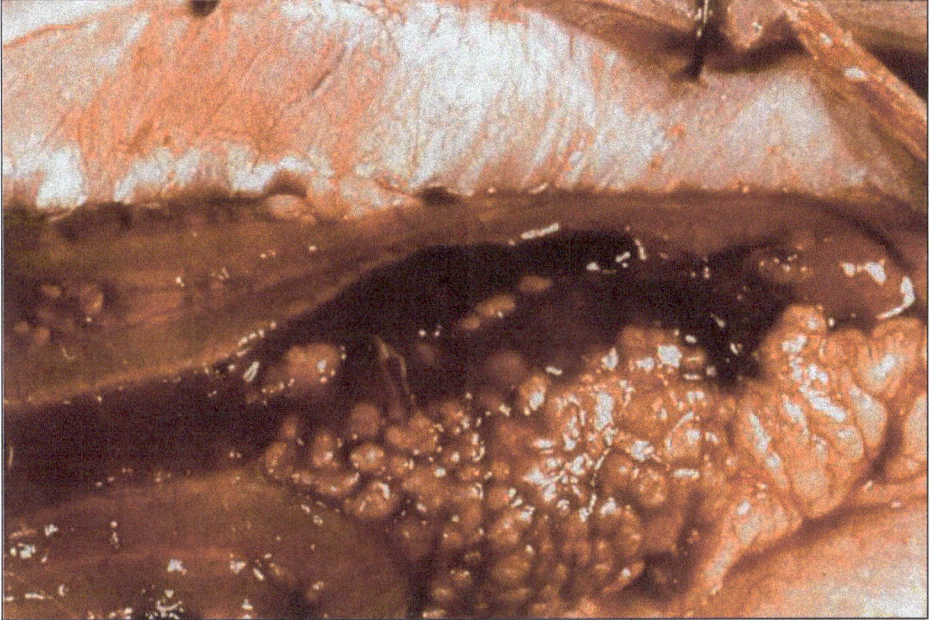

Figure 80: Granuloma in Liver of White Sucker.

References

Abdul Kader Mydeen, K.P. and Haniffa, M.A. (2011). Evaluation of antibacterial activity of medicinal plants on fish pathogen *Aeromonas hydrophila*. *J. Res. Biol.*, 1: 1-5.

Abowei, J.F.N. and Ekubo, A.T. (2011). A review of conventional and unconventional feeds in fish nutrition. *Br. J. Pharmacol. Toxicol.*, 2(4): 179-191.

Ahilan, B., Nithiyapriyatharshini, A. and Ravaneshwaran, K. (2010). Influence of certain herbal additives on the growth, survival and disease resistance of goldfish, *Carassius auratus* (Linnaeus). *Tamil Nadu J. Vet. Ani. Sci.*, 6(1): 5-11.

Ahmed, K., Akhand, A.A., Hasan, M., Islam, M. and Hasan, A. (2008). Toxicity of arsenic (sodium arsenite) to fresh water spotted snakehead *Channa punctatus* (Bloch) on cellular death and DNA content. *Am-Euras J. Agric. Environ. Sci.*, 4(1): 18-22.

Ainsworth, A.J., Dexiang, C. and Waterstrat, P.R. (1991). Changes in peripheral blood leukocyte percentages and function of neutrophils in stressed channel catfish. *J. Aquatic Anim. Health*, 3: 41-47.

Altinok, I., Capkin, E., Karahan, S. and Boran, M. (2006). Effects of water quality and fish size on toxicity of methiocarb, a carbamate pesticide, to rainbow trout. *Environ. Toxicol. Pharmacol.*, 22(1): 20-26.

Amanuma, K., Takeda, H., Amanuma. H and Aoki, Y. (2000). Transgenic zebrafish for detecting mutations caused by compounds in aquatic environments. *Nat. Biotechnol.*, 18: 62-65.

Azad, I.S., Dayal, J.S., Poornima, M. and Ali, S.A. (2007). Supra dietary levels of vitamins C and E enhance antibody production and immune memory in juvenile milkfish *Chanos chanos* (Forsskal) to formalin-killed *Vibrio vulnificus*. *Fish Shellfish Immunol.*, 23: 154-163.

Barbazuk, W.B., Korf, I., Kadavi, C., Heyen, J., Tate, S., Wun, E., Bedell, J.A., Mcpherson, J.A. and Johnson, S.L. (2000). The syntenic relationship of the zebrafish and human genomes. *Genome Res.*, 10: 1351-1358.

Bhat, B., Varma, A. and Saxena, G. (2012). Comparative study on the acute toxicity of plant based pesticide, kethrin and an organophosphate pesticide, dichlorvos to *Labeo rohita* (Hamilton). *Indian J. Life Sci.*, 2(1): 95-98.

Bielas, J.H. and Heddle, J.A. (2000). Proliferation is necessary for both repair and mutation in transgenic mouse cells. *Proc. Natl. Acad. Sci. U.S.A.*, 97: 11391-11396.

Bose, S., Nath, S. and Sahana, S.S. (2011). Toxic impact of thiamethoxam on the growth performance and liver protein concentration of a freshwater fish *Oreochromis niloticus* (Trewavas). *Ind. J. Fundamen. Appl. Life Sci.*, 1(4): 274-280.

Broughton, R.E., Milam, J.E. and Roe, B.A. (2001). The complete sequence of the zebrafish (*Danio rerio*) mitochondrial genome and evolutionary patterns in vertebrate mitochondrial DNA. *Genome Res.*, 11: 1958-1967.

Bunton, T.E. (1996). Experimental chemical carcinogenesis in fish. *Toxicol. Pathol.*, 24(5): 603-618.

Bunton, T.E. (1999). Use of non-mammalian species in bioassays for carcinogenicity. In: *Data on Genetic Effects in Carcinogenic Hazard Evaluation* (McGregor, D.B., Rice, J.M. and Venitt, S., eds.). IARC Scientific Publications, Lyon. pp. 151-184.

Burger, J. (2009). Risk to consumers from mercury in bluefish (*Pomatomuss altatrix*) from New Jersey: Size, season and geographical effects. *Environal Res.*, 109: 803-811.

Carvan, M.J., Dalton, T.P. and Stuart, G.W. (2000). Transgenic zebrafish as sentinels for aquatic pollution. *Ann. N.Y. Acad. Sci.*, 919: 133-147.

Castro, S.B.R., Leal, C.A.G., Freire, F.R., Carvalho, D.A., Oliveira, D.F. and Figueiredo, H.C.P. (2008). Antibacterial activity of plant extracts from Brazil against fish pathogenic bacteria. *Braz. J. Microbiol.*, 39(4): 756-760.

Chansue, N., Ponpornpisit, A., Endo, M., Sakai, M. and Satoshi, Y. (2000). Improved immunity of tilapia *Oreochromis niloticus* by C-UP III, a herb medicine. *Fish Pathol.*, 35: 89-90.

Chindah, A.C., Sikoki, F.D. and Vincent-Akpu, I. (2004). Toxicity of an organophosphate pesticide (chloropyrifos) on a common Niger Delta Wetland fish- *Tilapia guineensis* (Blecker 1862). *J. Appl. Sci. Environ. Manage.*, 8(2): 11-17.

Chitmanat, C., Tongdonmuan, K. and Nunsong, W. (2005). The use of crude extracts from traditional medicinal plants to eliminate *Trichodina* sp. in tilapia (*Oreochromis niloticus*) fingerlings. *Songklanakarin J. Sci. Technol.*, 27(Suppl 1): 359-364.

Christybapita, D., Divyagnaneswari, M. and Michael, R.D. (2007). Oral administration of *Eclipta alba* leaf aqueous extract enhances the non-specific immune responses and disease resistance of *Oreochromis mossambicus*. *Fish Shellfish Immunol.*, 23(4): 840-852.

Church, J.E. and Hodgson, W.C. (2002). The pharmacological activity of fish venoms. *Toxicon.*, 40(8): 1083-1093.

Clarkson, T.W. and Magos, L. (2006). The toxicology of mercury and its chemical compounds. *Critical Rev. Toxicol.*, 36(8): 609-662.

Cole, D.W., Cole, R., Gaydos, S.J., Gray, J., Hyland, G., Jacques, M.L., Powell-Dunford, N., Sawhney, C. and Au, W.W. (2009). Aquaculture: Environmental, toxicological and health issues. *Int. J. Hyg. Environ. Health*, 212: 369-377.

Daniels, R.J.R. (2002). *Freshwater Fishes of Peninsular India*. Universities Press (India), Hyderabad, AP, India.

Das, S. (2013). A review of dichlorvos toxicity in fish. Curr. *World Environ.*, 8(1): 143-149.

Deka, S. and Mahanta, R. (2012). A study on the effect of organophosphorus pesticide malathion on hepato-renal and reproductive organs of *Heteropneustes fossilis* (Bloch). *The Science Probe*, 1(1): 1-13.

Dhayanithi, N.B., Ajith Kumar, T.T. and Kathiresan, K. (2010). Effect of Neem extract against the bacteria isolated from marine fish. *J. Environ. Biol.*, 31: 409-412.

Dobaradaran, S., Naddafi, K., Nazmara, S. and Ghaedi, H. (2010). Heavy metals (Cd, Cu, Ni and Pb) content in two fish species of Persian Gulf in Bushehr Port. Iran. *Afr. J. Biotechnol.*, 9(37): 6191-6193.

Dugenci, S.K., Arda, N. and Candan, A. (2003). Some medicinal plants as immunostimulant for fish. *J. Ethanopharmacol.*, 88(1): 99-106.

Eggen, R.I.L., Behra, R., Burkhardt-Holm, P., Escher, B.I. and Schweigert, N. (2004). Challenges in ecotoxicology. *Environ. Sci. Technol.*, 38(3): 59A-64A.

Emami Khansari, F., Ghazi-Khansari, M. and Abdollahi, M. (2005). Heavy metals content of canned tuna fish. *Food Chem.*, 93: 293-296.

Engeszer, R.E., Patterson, L.B., Rao, A.A. and Parichy, D.M. (2007). Zebrafish in the wild: A review of natural history and new notes from the field. *Zebrafish*, 4(1): 21-40.

Fields, S. and Johnston, M. (2005). Cell Biology: Whither model organism research? *Science*, 307(5717): 1885-1886.

Fish Cancer Website (2015). www.google.com.

Fox, M.A. (1986). *The Case for Animal Experimentation: An Evolutionary and Ethical Perspective*. University of California Press, Berkeley and Los Angeles, California.

Gerhard, G.S., Kauffman, E.J., Wang, X., Stewart, R., Moore, J.L., Kasales, C.J., Demidenko, E. and Cheng, K.C. (2002). Life spans and senescent phenotypes of zebrafish (*Danio rerio*). *Exp. Gerontology*, 37: 1055-1068.

Ghazala, Mahboob, S., Al-Ghanim, K.A., Alkahem, Al-Balawi, H.F., Sultana, S., Sultana, T., Asi, M.R. and Ahmed, Z. (2014). Acute toxicity of commonly used organophosphates and carbamates to *Labeo rohita* fingerlings. *Bothalia Journal*, 44(1): 1-9.

Gilliom, R.J. (2007). Pesticides in US streams and groundwater. *Environ. Sci. Technol.*, 41(10): 3407-3413.

Gong, Z., Koh, C.H.V., Nguyen, A.T., Zhan, H., Li, Z., Lam, S.H., Spitsbergen, J.M., Emelyanov, A. and Parinov, S. (2011). The zebrafish model for liver carcinogenesis. *Cancer Genetics*, 5: 197-218.

Gupta, V. (2013). Mammalian feces as bio-indicator of heavy metal contamination in Bikaner Zoological Garden, Rajasthan, India. *Res. J. Animal, Veterinary and Fishery Sci.*, 1(5): 10-15.

Haitina, T., Klovins, J. and Schioth, H.B. (2005). Pharmacological characterization of melanocortin receptors in fish suggests an important role for ACTH. *Ann. New York Aca. Sci.*, 1040: 337-339.

Harikrishnan, R., Balasundaram, C. and Heo, M.S. (2010). Herbal supplementation diets on hematology and innate immunity in goldfish against *Aeromonas hydrophila*. *Fish Shellfish Immunol.*, 28(2): 354-361.

Harikrishnan, R., Nisha Rani, M. and Balasundaram, C. (2003). Hematological and biochemical parameters in common carp, *Cyprinus carpio*, following herbal treatment for *Aeromonas hydrophila* infection. *Aquaculture*, 221(1-4): 41-50.

Helena, M. and Zdenka, S. (2009). Incidence of animal poisoning cases in the Czech Republic: Current situation. *Interdisciplinary Toxicol.*, 2: 48-51.

Helfman, G., Collette Facey, D. and Bowen, B.W. (2009). *The Diversity of Fishes: Biology, Evolution, and Ecology.* Wiley-Blackwell, UK.

Hites, R.A., Foran, J.A., Carpenter, D.O., Coreen, H.M., Knuth, B.A. and Schwager, J.A. (2004). Global assessment of organic contaminants in farmed salmon. *Science*, 303(5655): 226-229.

Honda, S., Hylander, L. and Sakamoto, M. (2006). Recent advances in evaluation of health effects on mercury with special reference to methyl mercury: A mini review. *Environ. Hlth. Prev. Med.*, 11(4): 171-176.

Humphrey, H., Gardiner, J., Pandya, J., Sweeney, A., Gasior, D., McCaffrey, R. and Schantz, S. (2000). PCB congener profile in the serum of humans consuming Great Lakes fish. *Environ. Health Persp.*, 108(2): 167-72.

Ishikawa, T., Shimamine, T. and Takayama, S. (1975). Histologic and electron microscopy observations on diethylnitrosamine-induced hepatomas in small aquarium fish (*Oryzias latipes*). *J. Natl. Cancer Inst.*, 55: 909-916.

Jayakumari, Sahoo, P.K., Giri, S.S. and Pillai, B.R. (2004). Immunomodulation by ImmuPlus (AquaImmu) in giant freshwater prawn, *Macrobrachium rosenbergii* (De Man). *J. Exp. Bio.*, 42: 1073-1077.

Jeney, G., Galeotti, M., Volpatt, D., Jeney, Z. and Anderson, D.P. (1997). Prevention of stress in rainbow trout (*Oncorhynchus mykiss*) fed diets containing different doses of glucan. *Aquaculture*, 154: 1-15.

Johari, S.A., Kalbassi, M.R., Soltani, M, and Yu, I.J. (2013). Toxicity comparison of colloidal silver nanoparticles in various life stages of rainbow trout (*Oncorhynchus mykiss*). *Iranian J. Fishe. Sci.*, 12(1): 76-95.

Johnson, C. and Banerji, A. (2007). Influence of extract isolated from the plant *Sesuvium portulacastrum* on growth and metabolism in freshwater teleost, *Labeo rohita* (Rohu). *Fishery Technol.*, 44(2): 229-234.

Kent, M.L. and Bishop-Stewart, J.L. (2003). Transmission and tissue distribution of *Pseudoloma neurophilia* (Microsporidia) of zebrafish *Danio rerio*. *J. Fish Dis.*, 26: 1- 4.

Koenig, C.C. and Chasar, M.P. (1984). Usefulness of the hermaphroditic marine fish, *Rivulus marmoratus*, in carcinogenicity testing. *National Cancer Institute Monograph*, 65: 15-33.

Kumar, V., Abbas, A.K. and Fausto, N. (2006). *Pathologic Basis of Disease* (7th ed). Saunders, Elsevier India Pvt. Ltd., New Delhi. pp. 269-342 and 1059-117.

Kumari, M. (2012). Effects of organophosphate pesticide abate on the ovary of the cat fish, *Heteropneustes fossilis* (Bloch). *Bangladesh J. Zool.*, 40(2): 207-212.

Lakra, W.S., Swaminathan, T.R. and Joy, K.P. (2011). Development, characterization, conservation and storage of fish cell lines: A review. *Fish Physio. Bioch.*, 37(1): 1- 20.

Lamason, R.L., Mohideen, M.A. and Mest, J.R. (2005). SLC24A5, a putative cation exchanger, affects pigmentation in zebrafish and humans. *Science*, 310: 1782-1786.

Laetz, C.A., Baldwin, D.H., Collier, T.K., Hebert, V. Stark, J.D. and Scholz, N.L. (2009). The synergistic toxicity of pesticide mixtures: Implications for risk assessment and the conservation of endangered pacific salmon. *Environ. Health Persp.*, 117: 348-353.

Lee, J.S., Park, E.H, Choe, J. and Chipman, J.K. (2000). N-methyl-N-nitrosourea (MNU) induces papillary thyroid tumors which lack ras gene mutations in the hermaphroditic fish *Rivulus marmoratus*. Teratogen. *Carcinogen. Mutagen.*, 20: 1- 9.

Lee, S.J., Son, K.H., Chang, H.W., Kang, S.S. and Kim, H.P. (1998). Anti-inflammatory activity of *Lonicera japonica*. *Phytother. Res.*, 12: 445-447.

Legler, J., Broekhof, J.L.M., Brouwer, A., Lanser, P.H., Murk, A.J., van der Saag, P.T., Vethaak, A.D., Wester, P., Zivkovic, D. and van der Burg, B. (2000). A novel *in vivo* bioassay for (Xeno-) estrogens using transgenic zebrafish. *Environ. Sci. Technol.*, 34: 4439-4444.

Liu, D., Chen, W., Wei, J., Li, X., Wang, Z. and Jiang, X. (2012). A highly sensitive, dual-readout assay based on gold nanoparticles for organophosphorus and carbamate pesticides. American Chemical Society Publications.

Madhuri, S. (2012). Perspective on zebrafish: A suitable research model. *J. Exp. Zool. India*, 15(2): 303-307.

Madhuri, S. (2013). *A Manual on the Aquaculture Development of Freshwater Fish and Shellfish.* Narendra Publishing House, Delhi.

Madhuri, S. and Pandey, Govind (2009). Some anticancer medicinal plants of foreign origin. *Curr. Sci.,* 96(6): 779-783.

Madhuri, S. and Pandey, Govind (2012). Certain fish species used as models in cancer research. *Jigyasa,* 6(1): 65-70.

Madhuri, S. and Pandey, Govind (2014). *Toxicity and Treatment in Fish.* Narendra Publishing House, Delhi.

Madhuri, S., Mandloi, A.K., Pandey, Govind and Sahni, Y.P. (2012a). Antimicrobial activity of some medicinal plants against fish pathogens. *Int. Res. J. Pharm.,* 3(4): 28-30.

Madhuri, S., Mandloi, A.K., Pandey, Govind and Shrivastav, A.B. (2012b). Transgenic fish model in environmental toxicology. *Int. Res. J. Pharm.,* 3(5): 37-40.

Madhuri, S., Pandey, Govind, Bhandari, R. and Shrivastav, A.B. (2012c). Fish cancer developed by environmental pollutants. *Int. Res. J. Pharm.,* 3(10): 17-19.

Madhuri, S., Pandey, Govind and Mandloi, A.K. (2012d). Toxicity of certain pesticides in fishes. *Biochem. Cell. Arch.,* 12(2): 249-253.

Madhuri, S., Pandey, Govind and Sahni, Y.P. (2012e). Chemical toxicity leading to cancer in fishes and its treatment by medicinal plants. *Pl. Arch.,* 12(2): 579-584.

Madhuri, S., Sahni, Y.P., Mandloi, A.K. and Pandey, Govind (2012f). Toxicity in fish polluted with heavy metals, chemicals or drugs. *Jigyasa,* 6: 67-71.

Madhuri, S., Sahni, Y.P. and Pandey, Govind (2012g). Herbal feed supplements as drugs and growth promoter to fishes. *Int. Res. J. Pharm.,* 3(9): 30-33.

Madhuri, S., Shrivastav, A.B. and Pandey, Govind (2012h). Overviews of the zebrafish model and fish neoplasms. *The Global J. Pharmaceu. Res.,* 1(4): 736-743.

Madhuri, S., Shrivastav, A.B., Sahni, Y.P. and Pandey, Govind (2012i). Overviews of the treatment and control of common fish diseases. *Int. Res. J. Pharm.,* 3(7): 123-127.

Mahler, J.F. (2000). The use of genetically altered animals in toxicology. *Toxicol. Path.,* 28: 447-449.

Malla, F.A., Sharma, G. and Singh, S. (2009). Chlorpyrifos pesticide toxicity on erythrocyte sedimentation rate in fish, *Channa punctatus* (Bloch.). *Biol. Med.,* 1(2), 54-55.

MANAGE (2008). *Sustainable Fresh Water Aquaculture Development.* National Institute of Agricultural Extension Management. Hyderabad, AP, India.

Maniyar, R.A., Ahmed, R.N. and David, M. (2011). Monocrotophos: Toxicity evaluation and respiratory responses of *Cyprinus carpio* (Linnaeus). *Recent Res. Sci. Technol.,* 3(1): 51-54.

McCurry, J. (2006). Japan remembers Minamata. *Lancet,* 367(9505): 99-100.

Miklosi, A. and Andrew, R.J. (2006). The zebrafish as a model for behavioural studies. *Zebrafish*, 3: 227-234.

Mishra, B.P., Khan, M.Y., Badade, Z.G., Chandravathi, M. and Deepak (2013). Biochemical toxicity of organophosphate pesticide Metasystox in fish *Channa punctatus*. *Ind. J. of Forensic Med. Toxicol.*, 7(1): 254-287.

Mohammad Nejad Shamoushaki, M., Soltani, M., Kamali, A., Imanpoor, M.R., Sharifpour, I. and Khara, H. (2012). Effects of organophosphate, diazinon on some haematological and biochemical changes in *Rutilus frisii kutum* (Kamensky, 1901) male brood stocks. *Iranian J. Fish. Sci.*, 11(1): 105-117.

Montaser, M., Mahfouz, M.E., El-Shazly, S.A.M., Abdel-Rahman, G.H. and Bakry, S. (2010). Toxicity of heavy metals on fish at Jeddah coast KSA: Metallothionein expression as a biomarker and histopathological study on liver and gills. *World J. Fish Marine Sci.*, 2(3): 174-185.

Moore, M.J. and Stegeman, J.J. (1994). Hepatic neoplasms in winter flounder *Pleuronectes americanus* from Boston Harbor, Massachusetts, USA. Dis. Aquat. Org., 20: 33-48.

Mukherjee, S., Bidhan, C.K., Ray, S. and Ray, A. (2006). Environmental contaminants in pathogenesis of breast cancer. *Indian J. Exp. Biol.*, 44: 597-617.

Mukherjee, S.C. (2008). Anti-stress and adaptogenic effect of Aqua-CE on experimentally induced stress in Rohu (*Labeo rohita*). *Phytomedica*, 9: 1-3.

Myers, M.S., Stehr, C.M., Olson, O.P., Johnson, L.L., McCain, B.B., Chan, S.L. and Varanasi, U. (1994). Relationship between toxicopathic hepatic lesions and exposure to chemical contaminants in English sole *Pleuronectes vetulus*, starry flounder *Platichthys stellatus* and white croaker (*Genyonemus lineatus*) from selected marine sites on the Pacific coast, USA. *Environ. Health Persp.*, 102: 200-215.

Nargis, A., Khatun, M. and Talukder, D. (2011). Use of medicinal plants in the remedy of fish diseases. *Bangladesh Res. Publ. J.*, 5(3): 192-195.

Naveed, A., Venkaeshwarlu, P. and Janaiah, C. (2011). Biochemical alteration induced by triazophos in the blood plasma of fish, *Channa punctatus* (Bloch). *Ann. Biol. Res.*, 2(4), 31-37.

Norton, W.N. and Gardner, H.S. (2005). Diethylnitrosamine-induced spongiosis hepatis in medaka, *Oryzias latipes*. *Microsc. Microanal.*, 11(Suppl. 2): 1028-1029.

Nya, E.J. and Austin, B. (2009). Use of garlic, *Allium sativum*, to control *Aeromonas hydrophila* infection in rainbow trout, *Oncorhynchus mykiss* (Walbaum). *J. Fish Dis.*, 32(11): 963-970.

Okihiro, M.S. and Hinton, D.E. (1999). Progression of hepatic neoplasia in medaka (*Oryzias latipes*) exposed to diethylnitrosamine. *Carcinogenesis*, 20: 933-940.

Owena, S.F., Giltrowa, E., Huggett, D.B., Hutchinsonc, T.H., Sayed, J., Winterc, M.J. and Sumptera, J.P. (2007). Comparative physiology, pharmacology and toxicology of â-blockers: Mammals versus fish. *Aqu. Toxicol.*, 82(3): 145-162.

Pandey, A.K., Nagpure, N.S., Trivedi, S.P., Kumar, R., Kushwaha, B. and Lakra, W.S. (2011). Investigation on acute toxicity and behavioral changes in *Channa punctatus* (Bloch) due to organophosphate pesticide profenofos. *Drug and Chemical Toxicol.*, 34(4): 424-428.

Pandey, Ankita and Pandey, Govind (2013). Hazards of organophosphate and carbamate nanoparticles to fish species. *Int. J. Uni. Pharm. Biosci.*, 2(3): 17-27.

Pandey, Govind (2011a). A review of fish model in experimental pharmacology. *Int. Res. J. Pharm.*, 2(9): 33-36.

Pandey, Govind (2011b). Model organisms used in molecular biology or medical research. *Int. Res. J. Pharm.*, 2(11): 62-65.

Pandey, Govind (2013a). Treatment for certain parasitic diseases of fishes. *Univer. J. Pharm.*, 2(2): 1-3.

Pandey, Govind (2013b). Treatment and control of ornamental fish infection. *Univer. J. Pharm.*, 2(2): 52-56.

Pandey, Govind (2013c). Some medicinal plants to treat fish ectoparasitic infections. *Int. J. of Pharm. and Research Sci.*, 2(2): 532-538.

Pandey, Govind (2013d). Toxicity of cyanide in fishes: An overview. *Univer. J. Pharm.*, 2(2): 23-26.

Pandey, Govind (2013e). *Heavy Metals Poisoning in Fish*. International E - Publication, International Science Congress Association, Indore, MP, India.

Pandey, Govind (2013f). *Lab Manual of the Chemotherapy*. International E - Publication, International Science Congress Association, Indore, MP, India.

Pandey, Govind (2013g). Feed formulation and feeding technology for fishes. *Int. Res. J. Pharm.*, 4(3): 23-30.

Pandey, Govind (2013h). Overviews on diversity of fish. *Res. J. Animal, Veterinary and Fishery Sci.*, 1(8): 12-18.

Pandey, Govind (2014a). Potassium permanganate for the treatment of external infections of fish. *Jigyasa*, 8(1): 73-75.

Pandey, Govind (2014b). Ornamental fishes: Health managements by disinfectants and antibiotics. *Jigyasa*, 8(1): 60-66.

Pandey, Govind (2015). Herb-feed supplements for health and disease of fish. In: *Perspectives in Animal Ecology and Reproduction*, Vol. 10 (Gupta, V.K., Verma, A.K. and Singh, G.D., eds.). Daya Publishing House, New Delhi. pp. 171-182.

Pandey, Govind and Madhuri, S. (2010a). Significance of fruits and vegetables in malnutrition cancer. *Pl. Arch.*, 10(2): 517-522.

Pandey, Govind and Madhuri, S. (2010b). Cancer pathogenesis caused by xenoestrogens of environment and food contaminants: A review. *J. Chem. Pharm. Res.*, 2(4): 687-695.

Pandey, Govind and Madhuri, S. (2014). Heavy metals causing toxicity in animals and fishes. *Res. J. Animal, Veterinary and Fishery Sci.*, 2(2): 17-23.

Pandey, Govind, Madhuri, S. and Mandloi, A.K. (2012a). Medicinal plants useful in fish diseases. *Pl. Arch.*, 12(1): 1-4.

Pandey, Govind, Madhuri, S. and Mandloi, A.K. (2012b). Immunostimulant effect of medicinal plants on fish. *Int. Res. J. Pharm.*, 3(3): 112-114.

Pandey, Govind, Madhuri, S. and Sahni, Y.P. (2012c). Beneficial effects of certain herbal supplements on the health and disease resistance of fish. *Novel Sci.: Int. J. Pharmace. Sci.*, 1(7): 497-500.

Pandey, Govind, Madhuri, S. and Shrivastav, A.B. (2013a). Chemical carcinogenesis in fish: A review. *Univer. J. Pharm.*, 2(1): 14-20.

Pandey, Govind, Madhuri, S. and Shrivastav, A.B. (2013b). Prevalence of hepatic and skin cancers in fish by chemical contamination. *Int. J. of Pharm. and Research Sci.*, 2(2): 502-512.

Pandey, Govind, Madhuri, S. and Shrivastav, A.B. (2012d). Contamination of mercury in fish and its toxicity to both fish and humans: An overview. *Int. Res. J. Pharm.*, 3(11): 44-47.

Pandey, Govind, Madhuri, S. and Shrivastav, A.B. (2012e). Contamination of polychlorinated biphenyls in fishes leading to toxicity in fish consuming population. *Int. Res. J. Pharm.*, 3(12): 7-9.

Pandey, Govind, Madhuri, S. and Shrivastav, A.B. (2014). *Fish Cancer by Environmental Pollutants*, 1st Edn. Narendra Publishing House, Delhi.

Pandey, Govind, Shrivastav, A.B. and Madhuri, S. (2012f). Fishes of Madhya Pradesh with special reference to zebrafish as model organism in biomedical researches. *Int. Res. J. Pharm.*, 3(1): 120-123.

Pandey, Govind, Shrivastav, A.B. and Madhuri, S. (2015). *The Practical Book of Aquatic Animal Diseases and Health Cares*, 1st Edn. Narendra Publishing House, Delhi.

Patel, S., Gheewala, N., Suthar, A. and Shah, A. (2009). *In-vitro* cytotoxicity activity of Solanum nigrum extract against Hela cell line and Vero cell line. *Int. J. Pharm. Pharmaceu. Sci.*, 1(Suppl. 1): 8-10.

Poulet, F.M., Wolfe, M.J. and Spitsbergen, J.M. (1994). Naturally occurring orocutaneous papillomas and carcinomas of brown bullheads *Ictalurus nebulosus* in New York State. *Vet. Pathol.*, 31: 8-18.

Prashanth, M.S. (2003). Cypermenthrin induced physiological, biochemical and histopathological changes in freshwater fish, *Cirrhinus mrigala*. *PhD Thesis*. Karnataka University, Dharwad, India.

Prashanth, M.S. and Neelgund, S.E. (2008). Free cyanide induced biochemical changes in nitrogen metabolism of the Indian major carp, *Cirrhinus mrigala*. *J. Basic and Clin. Physiol. Pharmacol.*, 18(4): 277-287.

Prashanth, M.S. and Patil, Y.B. (2006). Behavioural surveillance of Indian major carp *Catla catla* (Hamilton) exposed to free cyanide. *J. Curr. Sci.*, 9(1): 313-318.

Prashanth, M.S., Sayeswara, H.A. and Goudar, M.A. (2011). Effect of sodium cyanide on behaviour and respiratory surveillance in freshwater fish, *Labeo rohita* (Hamilton). *Rec. Res. Sci. Tech.*, 3: 24-30.

Praveena, M., Sandeep, V., Kavitha, N. and Jayantha Rao, K. (2013). Impact of tannery effluent, chromium on hematological parameters in a fresh water fish, *Labeo rohita* (Hamilton). *Res. J. Animal, Veterinary and Fishery Sci.*, 1(6): 1-5.

Premdas, P.D. and Metcalfe, C.D. (1994). Regression, proliferation and development of lip papillomas in wild white suckers, *Catostomus commersoni*, held in the laboratory. *Environ. Biol. Fish.*, 40: 263-269.

Raisuddin, S. and Lee, J.S. (2008). Fish models in impact assessment of carcinogenic potential of environmental chemical pollutants: An appraisal of hermaphroditic mangrove killifish *Kryptolebias marmoratus*. In: *Interdisciplinary Studies on Environmental Chemistry- Biological Responses to Chemical Pollutants* (Murakami, Y., Nakayama, K., Kitamura, S-I., Iwata, H. and Tanabe, S., eds.). *Terrapub.* pp. 7-15.

Rajeshkumar, N.V., Pillai, M.R. and Kuttan, R. (2003). Induction of apoptosis in mouse and human carcinoma cell lines by *Emblica officinalis* polyphenols and its effect on chemical carcinogenesis. *J. Exp. Clin. Cancer Res.*, 22(2): 201-212.

Ravikumar, S., Anitha Anandha Gracelin, N., Palani Selvan, G. and Kalaiarasi, A. (2011). *In vitro* antibacterial activity of coastal medicinal plants against isolated bacterial fish pathogens. *Int. J. Pharmaceu. Res. Dev.*, 3(4): 109-116.

Ravikumar, S., Palani Selvan, G. and Anitha Anandha Gracelin, N. (2010). Antimicrobial activity of medicinal plants along Kanyakumari coast, Tamil Nadu, India. *Afr. J. Basic Appl. Sci.*, 2(5-6): 153-157.

Rodrigues, E. de L., Ranzani-Paiva, M.J.T., Pacheco, F.J. and Veiga, M.L. da (2001). Histopathologic lesions in the liver of *Prochilodus lineatus* (Pisces, Prochilodontidae) exposed to a sublethal concentration of the organophosphate insecticide Dipterex 500® (Trichlorfon). *Acta Scientiarum Maringa*, 23(2): 503-505.

Rubinstein, A.L. (2006). Zebrafish assays for drug toxicity screening. Expert Opinion on Drug Metabo. *Toxicol.*, 2: 231-240.

Sancho, E., Ferrando, M.D., Gamon, M. and Andreu-Moliner, E. (1992). Organophosphorus diazinon induced toxicity in the fish *Anguilla anguilla* L. *Compar. Biochem. Physiol. Part C: Compar. Pharmacol.* 103(2): 351-356.

Sastry, K.V. and Siddiqui, A.A. (1982). Chronic toxic effects of the carbamate pesticide sevin on carbohydrate metabolism in a freshwater snakehead fish, *Channa punctatus*. *Toxicol. Lett.*, 14(1-2): 123-130.

Satyavani, G., Gopi, R.A., Ayyappan, S., Balakrishnamurthy, P. and Reddy, N.P. (2011). Toxicity effect of expired pesticides to freshwater fish, *Labeo rohita*. *J. Agricul. Environ.*, 12: 1-8.

Sharma, A., Deo, A.D., Riteshkumar, S.T., Chanu, T.I. and Das, A. (2010). Effect of *Withania somnifera* (L. Dunal) root as a feed additive on immunological parameters and disease resistance to *Aeromonas hydrophila* in *Labeo rohita* (Hamilton) fingerlings. *Fish Shellfish Immunol.*, 29(3): 508-512.

Singh, R.N. (2013). Acute toxicity of an organophosphate, dimethoate to an air breathing fish, *Colisa fasciatus* (Bl. and Schn.). *Indian J. Sci. Res.*, 4(1): 97-100.

Singh, R.N., Pandey, R.K., Singh, N.N. and Das, V.K. (2009). Acute toxicity and behavioral responses of common carp *Cyprinus carpio* (Linn.) to an organophosphate (dimethoate). *World J. Zool.*, 4(2): 70-75.

Spence, R., Fatema, M.K., Ellis, S., Ahmed, Z.F. and Smith, C. (2007). The diet, growth and recruitment of wild zebrafish (*Danio rerio*) in Bangladesh. *J. Fish Biol.*, 71: 304-309.

Spence, R., Gerlach, G., Lawrence, C. and Smith, C. (2008). The behaviour and ecology of the zebrafish, *Danio rerio. Biol. Rev.*, 83: 13-34.

Srinivasarao, P., Bujjibabu, N. and Rameshraju, R. (2010). Study the effect of chlorpyrifos on proteins in fresh water fish *Labeo rohita* by using HPLC method. *Int. J. Res. Pharmace. Biomed. Sci.*, 1(1): 1-5.

Srivastava, A.K., Mishra, D., Shrivastava, S., Srivastav, S.K. and Srivastav, A.K. (2010). Acute toxicity and behavioural responses of *Heteropneustes fossilis* to an organophosphate insecticide, dimethoate. *Int. J. Pharma Bio Sci.*, 1(4): 359-363.

Stanton, M.F. (1966). Hepatic neoplasms of aquarium fish exposed to *Cycas cercinalis. Fed. Proc.*, 26: 661.

Surendra, K.Y. (2010). Pesticide applications: Threat to ecosystems. *J. Human Ecol.*, 32(1): 37-45.

Tatina, M., Bahmani, M., Soltani, M., Abtahi, B. and Gharibkhani, M. (2010). Effects of different levels of dietary vitamins C and E on some of hematological and biochemical parameters of starlet (*Acipenser ruthenus*). *J. Fish. Aqu. Sci.*, 5: 1-11.

Tilak, K.S., Mohanaranga Rao, D., Priyamvada Devi, A. and Murty, A.S. (1981). Toxicity of carbaryl and 1-naphthol to four species of freshwater fish. *J. Biosci.*, 3(4): 457-462.

Tilton, S.C., Gerwick, L.G., Hendricks, J.D., Rosato, C.S., Corley-Smith, G., Givan, S.A., Bailey, G.S., Bayne, C.J. and Williams, D.E. (2005). Use of a rainbow trout oligonucleotide microarray to determine transcriptional patterns in aflatoxin B1- induced hepatocellular carcinoma compared to adjacent liver. *Toxicol. Sci.*, 88: 319-330.

Turker, H., Yildirim, A.B. and Karakas, F.P. (2009). Sensitivity of bacteria isolated from fish to some medicinal plants. *Turkish J. Fish. Aqu. Sci.*, 9: 181-186.

Uppal, R.P. (2000). Toxicity of metals. In: *Veterinary Toxicology*, 1st Edn. (Garg, S.K., ed.). CBS Publishers and Distributers, New Delhi. pp. 37-59.

Vinodhini, R. and Narayanan, M. (2008). Bioaccumulation of heavy metals in organs of fresh water fish *Cyprinus carpio* (Common carp). *Int. J. Environ. Sci. Tech.*, 5(2): 179-182.

Wang, K., Wu, G. and Dai, S. (2003). Study on the immunological effect of the aqueous extract from Guangxi *Ganoderma leucidum* in the mice. *J. Guangxi Med. Univ.*, 871-874.

Webb, J.F. and Schilling, T.F. (2006). Zebrafish in comparative context: A symposium. *Integrat. Compar. Biol.*, 46: 569-576.

West, J., O'Neill, S., Lippert, G. and Quinnell, S. (2001). Toxic contaminants in marine and anadromous fishes from Puget Sound, Washington: Results of the Puget Sound assessment monitoring program fish component, 1989-1999. Technical Report FTP01-14, Washington Department of Fish and Wildlife, Olympia, WA.

White, R.M., Sessa, A. and Burke, C. (2008). Transparent adult zebrafish as a tool for in vivo transplantation analysis. *Stem Cell*, 2: 183-189.

Willemsen, R., Padje, S.V., van Swieten, J.C. and Oostra, B.A. (2011). Zebrafish (*Danio rerio*) as a model organism for dementia. *Neuromethods*, 48(4): 255-269.

Winkaler, E.U., Santos, T.R.M., Machado-Neto, J.G. and Martinez, C.B.R. (2007). Acute lethal and sub-lethal effects of neem leaf extract on the neotropical freshwater fish, *Prochilodus lineatus*. *Comp. Biochem. Physiol.* Part C, 145: 236-244.

Winn, R.N. (2001). Transgenic fish as models in environmental toxicology. *ILAR Journal*, 42(4): 1-6.

Yanong, R.P.E. (2014). Use of antibiotics in ornamental fish aquaculture. *Fisheries and Aquatic Sciences*, University of Florida.

Yin, G., Ardo, L., Jeney, Z., Xu, P. and Jeney, G. (2008). Chinese herbs (*Lonicera japonica* and *Ganoderma lucidum*) enhance non-specific immune response of tilapia, *Oreochromis niloticus*, and protection against *Aeromonas hydrophila*. In: *Diseases in Asian Aquaculture VI, Fish Health Section* (Bondad-Reantaso, M.G., Mohan, C.V., Crumlish, M. and Subasinghe, R.P., eds.). Asian Fisheries Society, Manila, Philippines. pp. 269-282.

Yue, X. and Dai, Z. (2011). Carbon nanotube-based cholinesterase biosensors for the detection of pesticides. In: *New Perspectives in Biosensors Technology and Applications* (Serra, P.A., ed.). pp. 403-420.

Zakia, A.M., Ahmed, M., Dosoki, I. and Shaimaa Abo, A. (2012). Occupational hazards in fish industry. *World J. Fish and Marine Sci.*, 4(2): 201-210.

Index

—

Colour Plates 209

Figure 4: Poisoned Fish. (Page 13)

Figure 5: Poisoned Fish. (Page 13)

Figure 6: Poisoned Fish. (Page 13)

Figure 7: Poisoned Fish. (Page 13)

Figure 8: Poisoned Fish. (Page 13)

Figure 17: Poisoned Fish. (Page 15)

Figure 18: Poisoned Fish. (Page 15)

Figure 76: Cell Tumour in Fish. (Page 185)

Figure 9: Poisoned Fish. (Page 14)

Figure 10: Poisoned Fish. (Page 14)

Figure 11: Poisoned Fish. (Page 14)

Figure 12: Poisoned Fish. (Page 14)

Figure 13: Poisoned Fish. (Page 14)

Figure 14: Poisoned Fish. (Page 14)

Figure 15: Poisoned Fish. (Page 14)

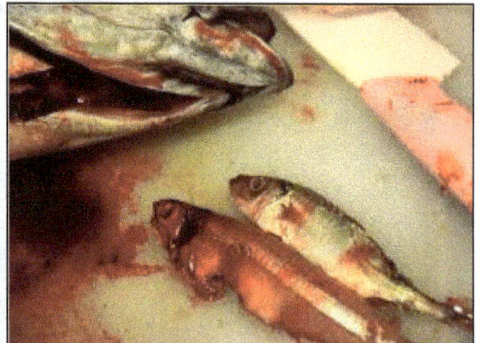

Figure 16: Poisoned Fish. (Page 14)

Figure 19: *Danio rerio.* (Page 26)

Figure 20: *Oncorhynchus mykiss.* (Page 26)

Figure 21: *Kryptolebias marmoratus.* (Page 26)

Figure 23: *Carassius auratus.* (Page 26)

Figure 22: *Oryzias latipes.* (Page 26)

Figure 24: *Cyprinus carpio.* (Page 26)

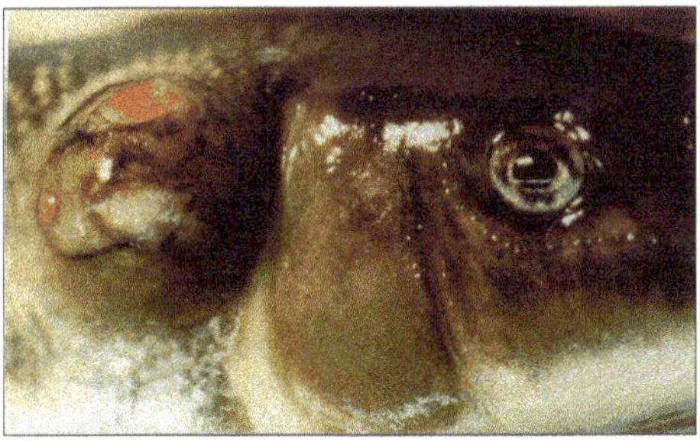

Figure 78: Fibroma in White Sucker. (Page 186)

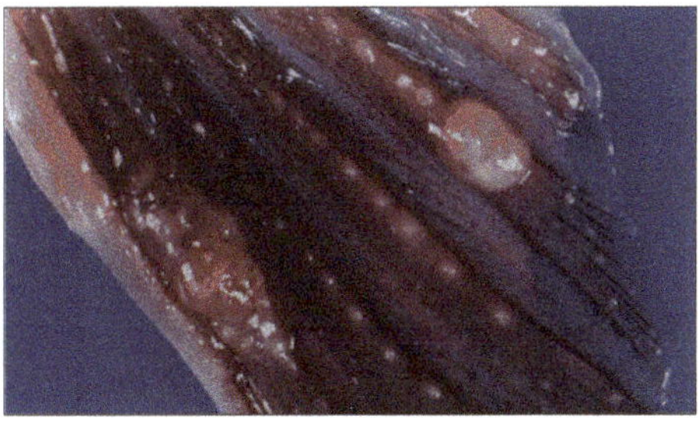

Figure 77: Papilloma on Fin of White Sucker. (Page 185)

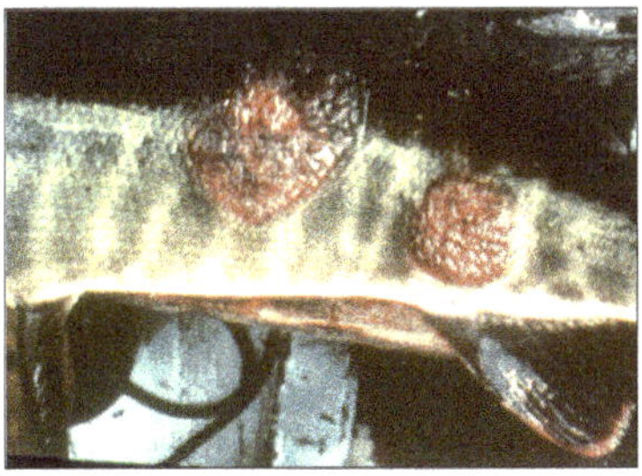

Figure 79: Lymphosarcoma in Muskellunge. (Page 186)

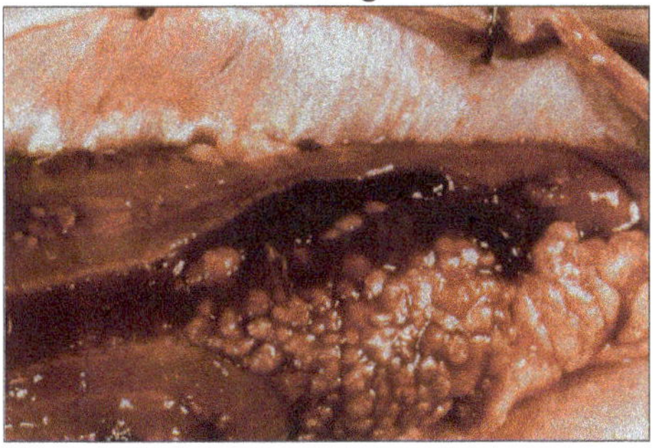

Figure 80: Granuloma in Liver of White Sucker. (Page 187)

Figure 25: *Poecilia reticulata.* **(Page 27)**

Figure 26: *Cyprinodon variegates.* **(Page 27)**

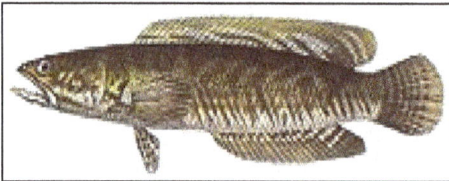

Figure 28: *Channa punctatus.* **(Page 28)**

Figure 27: *Catostomus commersoni.* **(Page 27)**

Figure 29: *Channa striata.* **(Page 28)**

Figure 30: *Labeo rohita.* **(Page 28)**

Figure 31: *Oreochromis niloticus.* **(Page 28)**

Figure 32: *Fundulus heteroclitus.* **(Page 28)**

Figure 33: *Catla catla.* **(Page 29)**

Figure 34: *Cirrhinus mrigala.* **(Page 29)**

Figure 35: *Oncorhynchus kisutch.* (Page 29)

Figure 36: *Tilapia guineensis.* (Page 30)

Figure 37: VHS with Telltale Haemorrhages (Red spots) on Bodies of Pacific Sardines. (Page 61)

Figure 38: IPN with Swollen Intestine of Salmon. (Page 62)

Figure 39: Columnaris Disease. (Page 64)

Figure 40: Dropsy in Goldfish. (Page 64)

Figure 41: Furunculosis in Salmon. (Page 67)

Figure 42: Mottled and Swollen Gills in BGD. (Page 67)

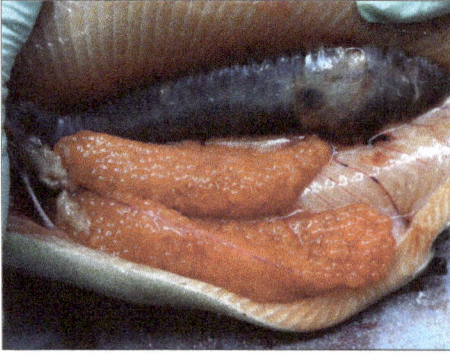

Figure 43: Bacterial Kidney Disease. (Page 68)

Figure 46: EUS Red Spots over Fish Body.
(Page 71)

Figure 48: Ick Disease in Fish (Page 73)

Figure 52: Velvet Disease. (Page 76)

Figure 44: Goldfish Ulcer Disease. (Page 68)

Figure 45: Saprolegniasis in Fish. (Page 70)

Figure 47: Branchiomycosis in Carp. (Page 71)

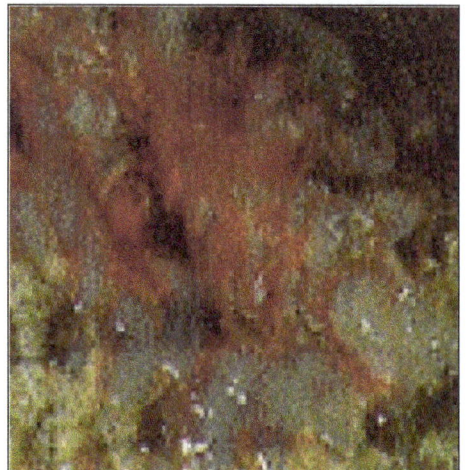

Figure 49: Trichodiniasis. (Page 74)

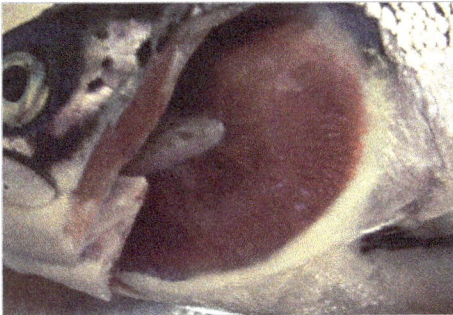

Figure 50: Amoebic Gill Disease in Salmon. (Page 75)

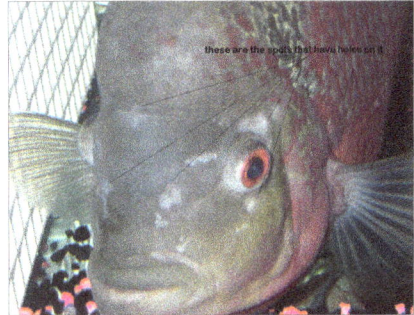

Figure 51: Hole in the Head Disease. (Page 75)

Figure 53: Neon Tetra Disease. (Page 77)

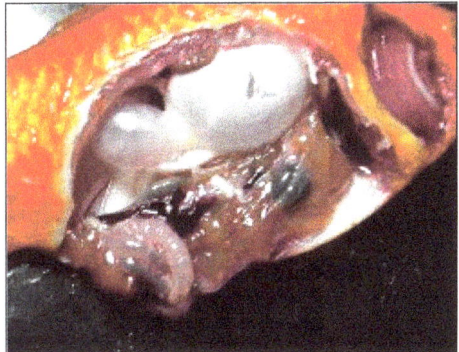

Figure 54: Swim Bladder Disease. (Page 78)

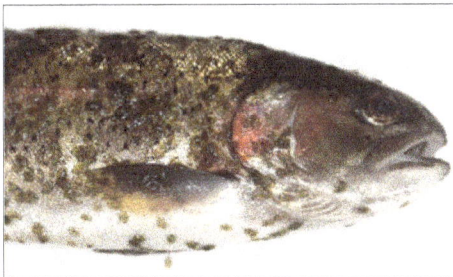

Figure 55: Argulus Infestation. (Page 79)

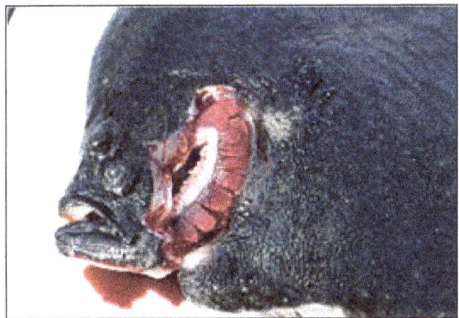

Figure 56: Gyrodactylosis on Skin of Atlantic Salmon. (Page 80)

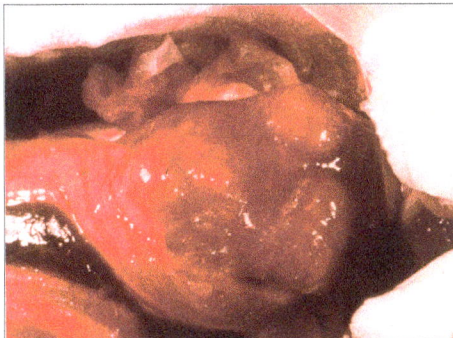

Figure 66: Cholangiolar and Hepatocellular Tumours in White Sucker. (Page 181)

Figure 67: Skin Cancer in Fish. (Page 182)

Figure 68: Skin Cancer in Walleye Fish. (Page 182)

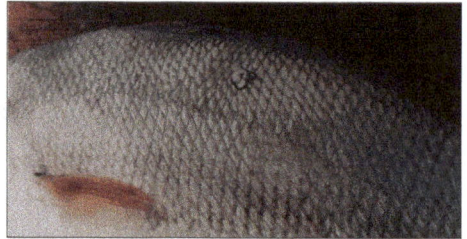

Figure 69: Dermal Cancer in Freshwater Drum. (Page 182)

Figure 70: Cancer on Head Region of Fish. (Page 183)

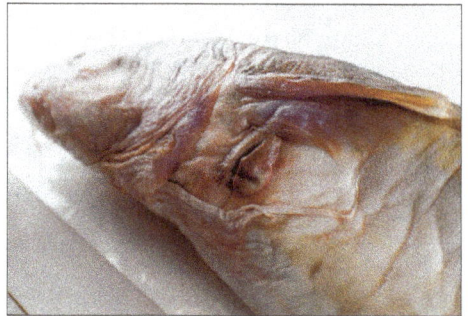

Figure 71: Cancer in Gills of Fish. (Page 183)

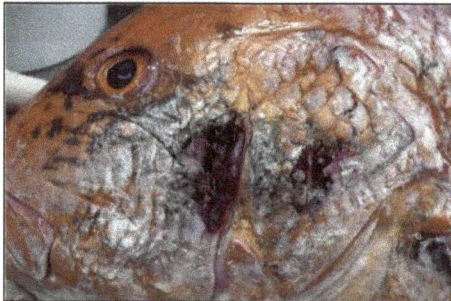

Figure 72: Cancer over Gill Region of Fish. (Page 184)

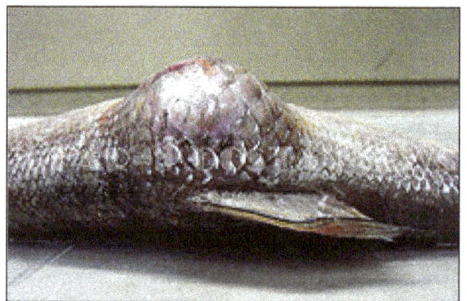

Figure 73: Tumour/Cancer in Fish. (Page 184)

Figure 74: Moth Cancer in Fish. (Page 184)

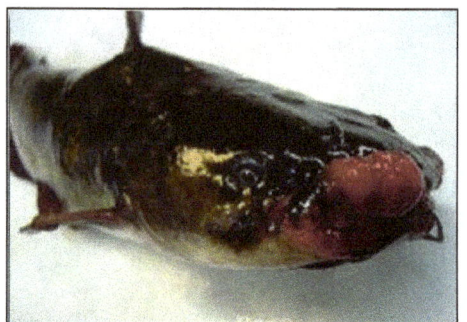

Figure 75: Cancerous Lesions in Mouth of Fish. (Page 185)

www.ingramcontent.com/pod-product-compliance
Lightning Source LLC
Chambersburg PA
CBHW050516190326
41458CB00005B/1560